Climate Change as a Threat to Peace

Dresdner Schriften zu Recht und Politik der Vereinten Nationen

Dresden Papers on Law and Policy of the United Nations

Herausgegben von/edited by
Sabine von Schorlemer

Zentrum für Internationale Studien/
School of International Studies,
Technische Universität Dresden/
University of Dresden

Band 19

PL ACADEMIC RESEARCH

Sabine von Schorlemer/Sylvia Maus (eds.)

Climate Change as a Threat to Peace

Impacts on Cultural Heritage and Cultural Diversity

PL ACADEMIC RESEARCH

Bibliographic Information published by the Deutsche Nationalbibliothek
The Deutsche Nationalbibliothek lists this publication in the Deutsche Nationalbibliografie; detailed bibliographic data is available in the internet at http://dnb.d-nb.de.

Library of Congress Cataloging-in-Publication Data
Climate change as a threat to peace : impacts on cultural heritage and cultural diversity / Sabine von Schorlemer, Sylvia Maus (eds.).
pages cm
"The contributions collected in this volume stem from an international conference entitled Climate Change as a Threat to Peace: Impacts on Cultural Heritage and Cultural Diversity, hosted by the German Commission for UNESCO and the UNESCO Chair in International Relations at the Technische Universität Dresden."
Includes bibliographical references.
ISBN 978-3-631-66223-6 – ISBN 978-3-653-05205-3 (e-book) 1. Climatic changes–Law and legislation–Congresses. 2. Environmental law, International–Congresses. 3. Climatic changes–Political aspects–Congresses. 4. Cultural property–Protection–Law and legislation–Congresses. 5. International law and human rights–Congresses. 6. Security, International–Congresses. I. Schorlemer, Sabine von, editor. II. Maus, Sylvia, 1982- editor.
K3585.5.A6C574 2015
363.738'74–dc23
2014044520

ISSN 1862-443X
ISBN 978-3-631-66223-6 (Print)
E-ISBN 978-3-653-05205-3 (E-Book)
DOI 10.3726/978-3-653-05205-3

PL Academic Research is an Imprint of Peter Lang GmbH.

Peter Lang – Frankfurt am Main · Bern · Bruxelles · New York · Oxford · Warszawa · Wien

This publication has been peer reviewed.

www.peterlang.com

Table of Contents

Lex Lata and _De Lege Ferenda_ – Legal Challenges of Cultural Property Protection in the Context of Climate Change

List of Abbreviations

CBD	Convention on Biological Diversity
CDG	Capacity Development and Governance
CDM	Clean Development Mechanism
CER	Certified Emission Reduction
COP	Conference of the Parties
CSICH	Convention for the Safeguarding of the Intangible Cultural Heritage
DESURBS	Designing Safer Urban Spaces
DRF	Dose-Response Functions
DRR	Disaster Risk Reduction
ECHR	European Court of Human Rights
ERU	Emission Reduction Unit
EU	European Union
FAO	Food and Agriculture Organisation
GHG	Green House Gas
HFA	Hyogo Framework of Action for Disaster Risk Reduction
HUL	Historic Urban Landscapes
ICCPR	International Covenant on Civil and Political Rights
ICCROM	International Centre for the Study of the Preservation and Restoration of Cultural Property
ICESCR	International Covenant on Economic, Social and Cultural Rights
ICJ	International Court of Justice
ICOMOS	International Council for Monuments and Sites
ICQHS	International Center on Qanats and Historic Hydraulic Structures
ICT	Information and Communications Technology
ILM	International Legal Materials
IPCC	Intergovernmental Panel on Climate Change
IUCN	International Union for Conservation of Nature
IWRM	Integrated Water Resources Management
LECZ	Low Elevation Coastal Zone
MAB	Man and the Biosphere Programme
NGO	Non-Governmental Organisation
OHCHR	Office of the High Commissioner for Human Rights
OUV	Outstanding Universal Value

OWHC	Organization of World Heritage Cities
SES	Social Ecological System
SFA	Systems and Flux Analysis
SLM	Soil and Land Use Management
UN	United Nations
UNEP	United Nations Environment Programme
UNESCO	United Nations Educational, Scientific and Cultural Organization
UNFCCC	United Nations Framework Convention on Climate Change
UNHRC	United Nations Human Rights Council
UNIDROIT	International Institute for the Unification of Private Law
UNISDR	United Nations Office for Disaster Risk Reduction
UNSC	United Nations Security Council
UNTS	United Nations Treaty Series
UNU	United Nations University
UNU-FLORES	United Nations University Institute for Integrated Management of Material Fluxes and of Resources
VCLT	Vienna Convention on the Law of Treaties
VMT	Vehicle Miles Travelled
WHC	World Heritage Convention
WHL	World Heritage List
WM	Waste Management
WRM	Water Resources Management

Sabine von Schorlemer & Sylvia Maus*

Reflections on Climate Change, Heritage and Peace

I Climate Change as a Threat to Peace

"Does Climate Change Kill People in Darfur?"[1] – The question is striking, but the title of a 2011 journal article captures the quintessence of a vivid and topical debate over the impacts of man-made climate change on international peace and security. Influential voices such as the UN Secretary-General[2] and the United Nations Environment Programme (UNEP)[3] have established a direct link between the armed conflict and climate change, notably for the conflict in Darfur.[4] A commentary of the International Institute for Strategic Studies partly attributes the Arab Spring to climate change.[5] And the well-known *Stern Review on the Economics of Climate Change* purports that "climate-related shocks have sparked violent conflict in the past".[6]

---

* Prof. Dr. jur. habil. Dr. rer. pol. habil. Sabine von Schorlemer is chair holder of the UNESCO Chair in International Relations at the Faculty of Law at the Technische Universität Dresden. On 30 September 2009, she was appointed as Saxon State Minister for Higher Education, Research and the Fine Arts.
Sylvia Maus, LL.M. (Nottingham) is a PhD candidate and scientific co-ordinator at the UNESCO Chair in International Relations.

1 Lyal S Sunga, 'Does Climate Change Kill People in Darfur?' (2011) 2(1) Journal of Human Rights and the Environment 64. See also Julie Flint and Alex de Waal, Darfur: A Short History of a Long War (Zed Books 2005).

2 Report of the Secretary-General on Climate Change and its Possible Security Implications, A/64/350 of 11 September 2009, especially paras. 68ff.

3 United Nations Environment Programme, Sudan: Post-Conflict Environmental Assessment (Nairobi 2007).

4 See e.g. Gareth Evans, Conflict Potential in a World of Climate Change, Address by the President of the International Crisis Group, to Bucerius Summer School on Global Governance 2008, Berlin, 29 August 2008, www.crisisgroup.org/en/publication-type/speeches/2008/conflict-potential-in-a-world-climate-change.aspx accessed 1 June 2014.

5 Sarah Johnstone and Jeffrey Mazo, 'Global Warming and the Arab Spring' (2011) 53(2) Survival: Global Politics and Strategy 11.

6 Nicholas Stern, The Economics of Climate Change (Cambridge University Press 2006), Executive Summary, p. vii-viii.

However, the patterns of correlation and causation are far from being firmly established and both critics and proponents of a conception of "climate change as conflict catalyst" are still wrestling with the problem.[7] While the link between scarcity of natural resources and conflict has long been established,[8] the claim that the "potential for the changing climate to induce conflict or exacerbate existing instability (…) is now recognized"[9] should be approached with caution. Critics argue that even though "there is unquestionably a general causal connection (...), at least in the sense that climate change is a 'threat multiplier'",[10] it is difficult to establish a clear causal link between climate change and conflict.[11] Indeed, in a long-term study ranging from 1990 to 2009, scholars examined climate-conflict relationships using data from over 16.000 violent events in East Africa (Burundi, Djibouti, Eritrea, Ethiopia, Kenya, Rwanda, Somalia, Tanzania, and

---

7 Idean Salehyan, 'From Climate Change to Conflict? No Consensus Yet' (2008) 45(3) Journal of Peace Research 315. For a summary of the debate see also Sumudu Atapattu, 'Climate Change, Resource Scarcity, Migration and Conflict: Implications for International Peace and Security' forthcoming in id. Human Rights Approaches to Climate Change: Challenges and Opportunities (Routledge, 2015) and, touching upon a variety of issues, Michael Brzoska, Martin Kalinowski, Volker Matthies and Berthold Meyer (eds) *Klimawandel und Konflikte. Versicherheitlichung versus Präventive Friedenspolitik?* (Nomos 2012).

8 See most notably Thomas F. Homer-Dixon, 'On the Threshold: Environmental Changes as a Cause of Acute Conflict' (1991) 16(2) International Security 76 and id., 'Environmental Scarcities and Violent Conflict: Evidence from Cases' (1994) 19:1 International Security 5.

9 Michael Werz and Laura Conley, 'Climate Change, Migration, and Conflict: Addressing Complex Crisis Scenarios in the 21st Century' (Center for American Progress and Heinrich-Böll-Stiftung 2013), www.boell.org/downloads/climate_migration_execsumm.pdf accessed 1 June 2014, 5. See further also Salehyan (n 7) and Attapatu (n 7).

10 Gareth Evans, 'Conflict Potential in a World of Climate Change', Address by the President of the International Crisis Group, to Bucerius Summer School on Global Governance 2008, Berlin, 29 August 2008, www.crisisgroup.org/en/publication-type/speeches/2008/conflict-potential-in-a-world-climate-change.aspx accessed 1 June 2014.

11 For studies questioning a causal link between climate change and conflict see, e.g. Buhaug, Halvard 'Climate not to blame for African civil wars' (2010) PNAS Early Edition, www.pnas.org/cgi/doi/10.1073/pnas.1005739107 accessed 1 June 2014; Tor A Benjaminsen, Koffi Alinon, Halvard Buhaug, Jill T Buseth 'Does climate change drive landuse conflicts in the Sahel?' (2012) 49 Journal of Peace Research 97; Erik, Gartzke 'Could climate change precipitate peace?' (2012) 49(1) Journal of Peace Research 177; Hanne Fjelde and Nina von Uexkull 'Climate triggers: Rainfall anomalies, vulnerability and communal conflict in Sub-Saharan Africa' (2012) 31(7) Political Geography 444.

Uganda)[12] and concluded that climate factors have a "modest influence in terms of predictive power in a model with political, economic, and physical geographic predictors".[13] They continue that "our findings question the most simplistic climate–conflict narratives. The relationships between rainfall and temperature variability and violence are complex and warrant careful interpretation".[14]

In 2007, the United Nations Security Council attended to the topic and held a high-level debate on climate change and international peace and security, addressing for the first time in this forum the potential impact of climate change on security.[15] In a *Statement by the President of the Security Council*, the "Security Council expresses its concern that possible adverse effects of climate change may, in the long run, aggravate certain existing threats to international peace and security".[16] The 2009 Secretary-General Report *Climate Change and its Possible Security Implications* provides a noteworthy summary of issues relating to climate change and security and identifies five "channels through which climate change could affect security":[17]

(a) Vulnerability: climate change threatens food security and human health, and increases human exposure to extreme events;
(b) Development: if climate change results in slowing down or reversing the development process, this will exacerbate vulnerability and could undermine the capacity of States to maintain stability;
(c) Coping and security: migration, competition over natural resources and other coping responses of households and communities faced with climate-related threats could increase the risk of domestic conflict as well as have international repercussions;

---

12 John O'Loughlina, Frank DW Witmer, Andrew M Linke, Arlene Laing, Andrew Gettelman, and Jimy Dudhia, 'Climate Variability and Conflict Risk in East Africa, 1990–2009' (2012) PNAS Early Edition, www.pnas.org/cgi/doi/10.1073/pnas.1205130109 accessed 1 June 2014.

13 Ibid, 1.

14 Ibid, 2.

15 UN Security Council, 17 April 2007, SC/9000, www.un.org/News/Press/docs/2007/sc9000.doc.htm accessed 1 June 2014.

16 UN Security Council, Statement by the President of the Security Council, 20 July 2011, S/PRST/2011/15. See also UN General Assembly Resolution 65/159 'Protection of Global Climate for Present and Future Generations of Humankind', 20 December 2010. For a summary of the activities within the UN, see Security Council Report, Monthly Forecast, July 2011, 30 June 2011, www.securitycouncilreport.org accessed 1 June 2014, 14.

17 UN General Assembly 'Climate Change and its Possible Security Implications' Report of the Secretary-General, 11 September 2009, A/64/350.

(d) Statelessness: there are implications for rights, security, and sovereignty of the loss of statehood because of the disappearance of territory;
(e) International conflict: there may be implications for international cooperation from climate change's impact on shared or undemarcated international resources.[18]

Equally, emerging threats which merit the attention of the international community are highlighted,[19] namely loss of territory, statelessness and increased numbers of displaced persons,[20] stress on shared international water resources,[21] and disputes surrounding the opening of the Arctic region to resource exploitation and trade.[22]

Overall, the report echoes the prevalent perception that conceives of climate change as a threat multiplier, "exacerbating threats caused by persistent poverty, weak institutions for resource management and conflict resolution, fault lines

---

18 Ibid, 1.

19 Ibid.

20 The issue of climate migrants or *climate refugees* has spurred considerable scholarly debate, see, for an early example, Norman Myers and Jennifer Kent, *Environmental Exodus: An Emergent Crisis in the Global Arena* (The Climate Institute 2005). See further Vikram O Kolmannskog, *Future Floods of Refugees: A Comment on Climate Change, Conflict and Forced Migration* (Norwegian Refugee Council 2008); Rosemary Rayfuse and Emily Crawford, 'Climate Change, Sovereignty and Statehood' (2011) Sydney Law School, Legal Studies Research Paper No. 11/59; Sumudu Atapattu, 'Climate Change: Disappearing States, Migration and Challenges for International Law' Washington Journal of Environmental Law & Policy (forthcoming). For a cautious view on a new treaty regime see, for instance, Jane McAdam, 'Swimming against the Tide: Why a Climate Change Displacement Treaty is Not the Answer' (2011) 23(1) International Journal of Refugee Law 2. For comprehensive reference to further literature, see Atapattu (n 7), note 64.

21 For instance, Nils P Gleditsch, Kathryn Furlong, Håvard Hegre, Bethany Lacina & Taylor Owen, 'Conflicts over Shared Rivers: Resource Scarcity or Fuzzy Boundaries?', (2006) 25(4) Political Geography 361; Ralf Ludwig, Roberto Roson, Christos Zografos and Giorgios Kallis, 'Towards an Interdisciplinary Research Agenda on Climate Change, Water and Security in Southern Europe and Neighboring Countries', (2011) 14 Environmental Science & Policy 794.

22 E.g. Susanne Wasum-Rainer, Ingo Winkelmann and Katrin Tiroch (eds) *Arctic Science, International Law and Climate Change: Legal Aspects of Marine Science in the Arctic Ocean* (Springer 2012); Robert Snyder, International Legal Regimes to Manage Indigenous Rights and Arctic Disputes from Climate Change (2011) 22 Colorado Journal of Int'l Environmental Law & Policy 1; Kathryn Isted, Sovereignty in the Arctic: An Analysis of Territorial Disputes and Environmental Policy Considerations (2009) 18(2) Journal of Transnat'l Law & Policy 343.

and a history of mistrust between communities and nations, and inadequate access to information or resources".[23]

II Impacts on Cultural Heritage and Cultural Diversity

In addition to these direct security implications, climate change is increasingly posing a threat to the protection of World Heritage. It affects cultural heritage (for example through temperature changes, soil erosion, flooding and storms) as well as natural heritage (e.g. through the melting of glaciers and habitat changes), posing a threat to biodiversity. Furthermore, climate change is likely to affect cultural diversity and socio-cultural interactions by forcing communities to change their work habits and ways of life, to compete for resources or to migrate elsewhere.[24]

The UNESCO mandate for the protection of cultural heritage and cultural diversity stems from four main conventions: First, the 1972 Convention concerning the Protection of the World Cultural and Natural Heritage,[25] with 190 States Parties, which links together the concepts of nature conservation and the preservation of cultural properties. The Convention sets out the duties of States Parties in identifying potential sites and their role in protecting and preserving them and focuses on the tangible aspects of heritage. The second Convention on the Protection of the Underwater Cultural Heritage,[26] adopted in 2001, with 42 States Parties today, outlines how States Parties should preserve underwater cultural heritage and take appropriate cooperative action. The third instrument

---

23 UN General Assembly 'Climate Change and its Possible Security Implications' Report of the Secretary-General, 11 September 2009, A/64/350, p. 1.

24 UNESCO World Heritage Centre, 'Climate Change and World Heritage. Report on predicting and managing the impacts of climate change on World Heritage and Strategy to assist States Parties to implement appropriate management responses' World Heritage Reports No 22, May 2007, available at whc.unesco.org/documents/publi_wh_papers_22_en.pdf. For Case Studies and general comments on the management of world heritage sites in the context of climate change see UNESCO World Heritage Centre (2007a) 'Case Studies on Climate Change and World Heritage', (June 2007, re-issued May 2009), available at http://whc.unesco.org/uploads/activities/documents/activity-43-9.pdf and the special issue on climate change of the quarterly journal "World Heritage No. 42" (June 2006).

25 Convention concerning the Protection of the World Cultural and Natural Heritage; adopted 16 November 1972, entered into force 17 December 1975, 1037 UNTS 151.

26 Convention on the Protection of the Underwater Cultural Heritage; adopted 2 November 2001, entered into force 2 January 2009, (2002) 41 ILM 37.

is the 2003 Convention for the Safeguarding of Intangible Cultural Heritage[27] with currently 153 States Parties. This Convention highlights that cultural heritage does not end at monuments and collections of objects, but also includes oral traditions, performing arts, and social practices, inherited from our ancestors. The fourth convention of importance is the 2005 Convention on the Protection and Promotion of the Diversity of Cultural Expressions,[28] with 127 States Parties and the European Union, ensuring that artists, cultural professionals, practitioners and citizens worldwide can create, produce, disseminate and enjoy a broad range of cultural goods, services and activities, including their own.[29] As a fifth relevant legal instrument, the Convention on the Protection of Cultural Property in the Event of Armed Conflict from 1954,[30] including its Second Protocol from 1999,[31] could be added to the aforementioned list of UNESCO conventions, with regards to the special circumstances under international humanitarian law. While these conventions cover a range of topics, climate change is not particularly mentioned in any of them. A reference to *climatic factors* can only be found in the Operational Guidelines for the Implementation of the World Heritage Convention[32]; first, as "factors affecting the property" in the Nomination Format for inscription on the World Heritage List (Annex 5, 4.b (ii)) and, second, as

27 Convention for the Safeguarding of the Intangible Cultural Property; adopted 17 October 2003, entered into force 20 April 2006, 2368 UNTS 3.

28 Convention on the Protection and Promotion of the Diversity of Cultural Expressions; adopted 20 October 2005, entered into force 18 March 2007, 2440 UNTS 311.

29 Sabine von Schorlemer: Kulturpolitik im Völkerrecht verankert. Das neuen UNESCOÜbereinkommen zum Schutz der kulturellen Vielfalt, (2005) 6 Zeitschrift Vereinte Nationen 217; Sabine von Schorlemer and Peter-Tobias Stoll (eds.), *The UNESCO Convention on the Protection and Promotion of the Diversity of Cultural Expressions. Explanatory Notes* (Springer 2012).

30 Convention for the Protection of Cultural Property in the Event of Armed Conflict with Regulations for the Execution of the Convention; adopted 14 May 1954, entered into force 7 August 1956, 249 UNTS 240.

31 Second Protocol of 1999 to the Hague Convention of 1954 for the Protection of Cultural Property in the Event of Armed Conflict; adopted 26 March 1999, entered into force 9 March 2004, 2253 UNTS 172.

32 The Operational Guidelines for the Implementation of the World Heritage Convention are drafted and regularly updated by the World Heritage Committee. They contain precise criteria for the implementation of the Convention, including for the inscription of properties on the World Heritage List and the List of World Heritage in Danger, see whc.unesco.org/en/guidelines.

"potential danger" leading to the inscription of properties on the List of World Heritage in Danger (Paragraphs 179b (vi)).

The impacts on climate change on cultural and natural heritage were put on the agenda of the UNESCO World Heritage Committee for the first time in 2005. The Committee concluded that "the impacts of climate change are affecting many and are likely to affect many more World Heritage properties, both natural and cultural in the years to come".[33] One year later, the group of experts consisting of representatives of the World Heritage Centre and of the advisory bodies,[34] submitted its report "Predicting and Managing the Effects of Climate Change on World Heritage" as well as the "Strategy to assist States Parties to implement appropriate management responses".[35] Both documents present a detailed analysis of the threats faced by World Heritage sites and discuss potential preventive and corrective measures. Last but not least, the World Heritage Committee in 2007 adopted the "Policy Document on the Impacts of Climate Change on World Heritage Properties",[36] identifying synergies between existing international legal instruments as well as legal questions concerning the role of the World Heritage Convention with respect to an appropriate reaction to climate change. As the Policy Document highlights, climate change poses a variety of questions, not least concerning the legal obligations of Members to the World Heritage Convention and institutional challenges with respect to climate change.[37] Rather surprisingly, there seemingly was no follow-up on these documents,[38] except for an amendment of the Operational Guidelines,[39] which now acknowledge the "threatening

33 UNESCO-Document 29COM 7B.a, 2005, Nr. 5.

34 International Union for Conservation of Nature (IUCN), International Council on Monuments and Sites (ICOMOS), International Centre for the Study of the Preservation and Restoration of Cultural Property (ICCROM).

35 WHC-06/30.COM/7.1, 2006.

36 WHC-07/16.GA/10, 2007.

37 Ibid. For a discussion of the scope of legal obligations under the Convention with respect to climate change, see Sylvia Maus, 'Hand in Hand Against Climate Change: Cultural Human Rights and the Protection of Cultural Heritage', Cambridge Review of International Affairs (forthcoming 2014).

38 As the latest issue of the World Heritage Series, UNESCO published a management tool intended at assisting site managers in their responses to climate change, however without further taking up broader issues of World Heritage and climate change, see Jim Perry and Chalie Falzon, 'Climate Change Adaptation for Natural World Heritage Sites. A Practical Guide' (2014) UNESCO World Heritage Series No 37.

39 UNESCO, World Heritage Committee Decision 32 COM 7A.32 (31 March 2009) WHC08/32.COM/24Rev para 5.

impact of climatic (…) factors"[40] as one of the threats having potentially deleterious effects on the inherent characteristics of a property. A more determined reaction by UNESCO to climate change is wanting. For instance, no further activities such as further amendment of the Operational Guidelines to better account for the threats imposed by climate change or the adoption of an additional protocol on this topic have taken place to date.[41] Similarly, existing tools, most notably the Fund for the Protection of the World Cultural and Natural Heritage of Outstanding Universal Value,[42] bear the potential of playing a crucial important role in protecting properties threatened by climate change.[43] While bearing in mind its limited size, this potential has to date not received much attention. As an international organisation, UNESCO also faces institutional challenges in the context of climate change that call for closer co-operation between the different Programme Sectors *Education*, *Sciences*, *Culture* and *Communication and Information*. Furthermore, enhanced exchange between UNESCO and other relevant actors in the field such as States Parties to the UNESCO conventions, but also related UN bodies and agencies, other international organisations, and national stakeholders would be essential. Yet, instead of assuming a leading role,[44] UNESCO and the World Heritage Centre willingly left the field to the United Nations Framework Convention on Climate Change (UNFCCC) by stating that the "UNFCCC is the UN instrument through which mitigation strategies at the global and States Parties level is being addressed".[45]

III Framework of Analysis

The present volume takes a fresh look at the issues and unanswered questions presented above and it intends to fill some of the gaps which have been identified. The contributions collected in this volume stem from an international conference

40 Paragraph 179(b)(vi) of the Operational Guidelines. See also the contributions of Federico Lenzerini and of Alessandro Chechi in this volume.

41 See in detail the contribution of Alessandro Chechi in this volume.

42 Article 15 WHC.

43 See Federico Lenzerini, 'Articles 15-16 World Heritage Fund', in Francesco Francioni with Federico Lenzerini (eds.), *The 1972 World Heritage Convention. A Commentary* (Oxford University Press 2008), 269-288.

44 On the potential of UNESCO as a leading organisation, see Martin Gerner, *Lead Agency: UNESCO's Global Leadership and Co-ordination Role for the United Nations Decade of Education for Sustainable Development (2005-2014)*, Dresden Papers on Law and Policy of the United Nations Vol 17 (Peter Lang 2013).

45 (n 35), para 18.

entitled *Climate Change as a Threat to Peace: Impacts on Cultural Heritage and Cultural Diversity*, hosted by the German Commission for UNESCO and the UNESCO Chair in International Relations at the Technische Universität Dresden. It proceeded under the assumption that the impacts of climate change on cultural heritage and cultural diversity may challenge sustainable global peace. Accordingly, the above mentioned threats on climate change and conflict on the one hand, and protection of cultural property and climate change on the other one, were pulled together into one conceptual triangle. Maybe as the most innovative feature, the nexus between cultural heritage and peace was explicitly taken account of. The UNESCO Chair in International Relations dedicates much of its research to the relationship between peace and cultural heritage and cultural diversity. While the importance of the protection of cultural heritage in armed conflicts tends to become more and more recognized,[46] the crucial role of cultural policy as a reconciliatory, proactive element of building and securing of sustainable peace[47] has so far been largely underestimated. Linking this nexus to the implications of climate change on both cultural property and peace is highly topical and thus almost a self-evident continuation of the research at the UNESCO Chair in International Relations.

To be sure, the described triangular integration is a complex one, one that needs innovative thinking and innovative action. It is also one that has to be acknowledged as an interdisciplinary issue bringing together natural sciences and

---

46 E.g. Sabine von Schorlemer, 'Cultural Heritage Law: Recent Developments in the Laws of War and Occupation' in James AR Nafziger and Ann M Nicgorski, *Cultural Heritage Issues: The Legacy of Conquest, Colonization and Commerce* (Martinus Nijhoff 2010) 137-158. Currently, the UNECO Chair in International Relations conducts a research project on the "menace to cultural property in armed conflicts" in which the mechanisms for protecting cultural properties and cultural heritage in armed conflicts, with special emphasis on the provisions of the relevant conventions, are analysed. Furthermore, their application and effectiveness in the light of recent challenges illustrated by current examples of armed conflicts (amongst others Thailand und Cambodia, Afghanistan, Iraq, Congo, Egypt, Tunisia, Libya, Syria, Mali) are examined.

47 See, as an early recognition of this connection, Walter Hirche, Kulturelle Zusammenarbeit als dritte Dimension der deutschen Außenpolitik, in: Deutsche UNESCO-Kommission (Hrsg.), Lernziel Weltoffenheit, Fünfzig Jahre deutsche Mitarbeit in der UNESCO (2001), p. 31, who regards cultural cooperation, especially in order to protect cultural heritage and diversity, by its very nature as a *trust building task* bearing, as a matter of principle, the potential for "endowing legal relationships with respect to peaceful cooperation" (translation by the authors).

engineering with humanities and social sciences. The focus of the conference was on the legal and institutional challenges faced by national and international actors, and especially by UNESCO. However, as the present contributions show, it also alluded to broader issues of mitigation, adaptation and resilience.

The first part of the volume, *Interdisciplinary Approaches of Linking Climate Change with Politics and Law of Cultural Heritage*, starts off with shorter articles and comments from both academics and practitioners and intends to map the field and set the scene. In his contribution *Man-made Climate Change: A major Challenge for World Heritage Conservation*, *Bernd von Droste zu Hülshoff*, former Assistant Director General of UNESCO and Founding Director of the UNESCO World Heritage Centre, gives an overview about the impacts of climate change on natural and cultural heritage as well as on past, current and future activities of UNESCO and the World Heritage Committee in particular. As an opening of the debate, *von Droste zu Hülshoff* underlines the special obligation of rich countries to "commit to protecting our common global heritage" and calls for enforceable international obligations of "caring for the world climate" and "caring for our cultural and natural inheritance".

Gunter Pleuger, long-time diplomat and Permanent German Representative at the UN in New York from 2002-2006, analyses the role of the UN Security Council in his article *Climate Change as a Threat to International Peace – The Role of the UN Security Council*. While acknowledging the capacity of the Security Council to counter the lack of legal and enforceable obligations by taking binding decisions, and even "establishing itself as a new source of international law next to international customary and treaty law", *Pleuger* does not see the issue of climate change as a threat to peace to be placed on the Security Council agenda in the imminent.

With her contribution *The United Nations Human Rights-Based Approach to Climate Change – Introducing a Human Dimension to International Climate Law*, *Franziska Knur* article revisits the 2009 Office of the High Commissioner for Human Rights Report on the relationship of climate change and human rights and undertakes to develop the United Nations human rights-based approach to climate change into a more meaningful concept. She comes to the conclusion that a human rights-based approach can have significant impact on the international climate change discourse, not least by providing access to institutions for the concretization of State obligations and by the introduction of a vertical dimension, *id est* subjective rights, into international climate law.

The second part of the book illuminates in some details the manifold and sometimes severe *Impacts of Climate Change on Cultural Heritage and on Cultural Diversity*. *Michael Turner* and *Rachel Singer* focus, from a perspective of Urban Design and Architecture, on the role of *Urban Resilience in Climate Change*. Treating the concept of *resilience* as a "panacea to the negative impacts of change", *Turner* and *Singer* propose new integrative conceptual tools to understand Urban Resilience and stress the importance of heritage and cultural diversity in the discussion in order to ensure sustainability between communities and preserving heritage for future generations.

Roger-Alexandre Lefèvre skilfully addresses the distinction between the *Impact of Climate Change on Slow Degradation of Monuments in Contrast to Extreme Events*. While the focus of attention in the realm of climate change often lies with extreme weather events such as extreme winds, storms, tornados; extreme precipitations, flooding, flash floods; heat waves, drought; pollution peaks, built cultural heritage will also experience slow events in strong correlation with air pollution (e.g. recession of façades in limestone or marble; soiling of stone surface and of glass; chemical leaching of mediaeval stained glass; metal corrosion) or weak correlation with air pollution (salt crystallization in porous walls; freeze-thaw damage in porous materials; submersion of monuments on littoral due to sea level rise; swelling-shrinkage of expansive clay minerals in soils; biomass accumulation on façades in unurban areas). In conclusion of his contribution, *Lefèvre* poses the question as to whether monuments will really be threatened by these slow climate change related events. Interestingly, he comes to the conclusion that while monuments in Central and Southern Europe are not likely to be threatened by the slow evolution, monuments in Northern Europe will be threatened. Therefore, *Lefèvre* calls for improvement basic research in order to better anticipating climate change impacts on cultural heritage.

With their contribution *Impact of Global Change on World Heritage and on Environmental Resources: The Need for an Integrated Management Approach*, *Reza Ardakanian* and *Stephan Hülsmann* consider heritage sites within their broader natural environment and the corresponding environmental resources. They focus on the need for an integrated approach for mitigating and/or adapting to the direct and indirect impacts of global climate change and argue that both protection and management of heritage sites must be approached by considering the environmental resources within the respective region. Especially with regards to developing countries, *Ardakanian* and *Hülsmann* call for management strategies to be embedded in capacity development activities addressing all relevant stakeholders.

The contribution of *Elisabeth Longworth The Culture of Prevention: Heritage and Resilience* focuses on the important role cultural heritage plays in supporting sustainable development, disaster risk reduction and building community resilience. Against this backdrop, *Longworth* calls for enhanced action to be taken in order to support efforts towards heritage-driven resilience, *inter alia* through the integration of heritage in disaster risk reduction plans, policies and strategies at all levels as well as capacity-building programmes and awareness raising. Notably, such enhanced action does not only include cooperation between different state institutions but also an inclusion of the private sector by harnessing public-private partnerships.

The contributions of the next part entitled *Lex Lata and De Lege Ferenda – Legal Challenges of Cultural Property Protection in the Context of Climate Change* address the legal challenges of protecting cultural property in the face of climate change. In the focus of attention are amongst others the potential consequences of the failure of a State to protect cultural property in the context of climate change as well as further breaches of international obligations. Existing legal rules are examined and legal voids identified which may call for amendments to prevailing treaty law or even the drafting of a new international legal instrument.

The relationship between cultural heritage protection and climate change is graphically presented in *Guido Carducci's* contribution *What Consideration is Given to Climate and to Climate Change in the UNESCO Cultural Heritage and Property Conventions*? Notably, *Carducci* elaborates on the possibilities but also the limits of law and legal measures in this regard before giving an overview of UNESCO's legal instruments protecting tangible and intangible heritage and how they treat (or rather do not treat) climate change.

Federico Lenzerini sheds light to some of these issues with his contribution *Protecting the Tangible, Safeguarding the Intangible: A Same Conventional Model for Different Needs*. While acknowledging the importance and the success of the 1972 World Heritage Convention, *Lenzerini* is more reluctant with the replication of this model in other legal instruments, most notably in the 2003 Convention for the Safeguarding of the Intangible Cultural Heritage. With respect to climate change, *Lenzerini* explains that both Conventions "in principle provide the necessary tools to face the threats posed by climate change". However, practice shows a lack of adequate attention to the issue within both legal instruments. While the WHC at least mentions climate change in its Operations Guidelines, the CSICH so far lacks such references in its Operational Directives. In conclusion, *Lenzerini* concedes that any changes to the current situation depend on the political will of the respective states parties. However, it is warned that the "international community cannot further postpone setting up a global, targeted and efficient action having the purpose of fighting the detrimental effects produced by climate change on both tangible and intangible cultural heritage. Maybe the time is still right to do that; the risk is that very soon it may be too late".

Subsequently, *Alessandro Chechi* analyses the *The Cultural Dimension of Climate Change: Some Remarks on the Interface between Cultural Heritage and Climate Change Law*. By examining the existing international legal framework, he investigates whether and to what extent cultural heritage law interacts with the international legal climate change regime and argues that the recognition of cultural heritage concerns by climate change law is both desirable and necessary. Notably, while, in *Chechi's* view, the WHC does contain a legally binding obligation for the States Parties to adopt GHG emissions limits "more compelling than those called for by the Kyoto Protocol", he also draws attention to other mechanisms such as the Fund and the Operational Guidelines. Furthermore, existing shortcomings in the climate change regime must be addressed, bearing in mind that "procrastination is not an option".

Birgitta Ringbeck concludes this section with her chapter on *World Cultural Heritage Sites and Climate Change: Management Issues*. By calling for a full integration of heritage concerns into risk strategies and programs, *Ringbeck* underlines the importance of management plans for World Heritage sites as "keytools in the effective stewardship of World Heritage sites under threat from climate change". In addition, general mitigation policy is needed, such as the expansion of renewable energy, which has been promoted by German national and federal governments. In doing so, however, *Ringbeck* regrets that not enough attention is given to the potential of the use and reuse of moments and their appreciation as non-renewable resources.

As both a summary and outlook, *Roland Bernecker's Concluding Remarks* recapitulate the main findings of the foregoing contributions. In bringing together the three elements of cultural heritage, climate change and peace, *Bernecker* once more underlines that culture is a "crucial, yet dramatically neglected dimension of sustainability". Furthermore, he encourages scholars, practitioners and other stakeholders to broaden their views and rework familiar concepts such as preservation, safeguarding, change, and management. Especially the concept of resilience has emerged as crucial in this debate. As a resumé of the challenges we are facing today, *Bernecker* appositely concludes that "our inventiveness and our hunger for improvement not only compromise our future, but consume our past". In order to prevent this from happening, a return to the universal values highlighted most notably by the World Heritage Convention, but also by her sister instruments, is "a necessary contribution to peace".

The conference *Climate Change as a Threat to Peace: Impacts on Cultural Heritage and Cultural Diversity* and thus also this volume could only be realised thanks to the generous financial contributions of the Association of Friends and Sponsors of the Technische Universität Dresden (Gesellschaft von Freunden und

Förderern der TU Dresden), the School of International Studies (Zentrum für Internationale Studien, ZIS) and the Regional Association Saxony, SaxonyAnhalt and Thuringia of the United Nations Association of Germany (Landesverband Sachsen, Sachsen-Anhalt und Thüringen der Deutschen Gesellschaft für die Vereinten Nationen e.V., DGVN). The editor wishes to express deep gratitude to these organisations for their support. For their excellent work and commitment thanks are extended to the organising committee of the conference as well as the editorial team at the UNESCO Chair: Franziska Knur, Daphne Rodzinyak, Antje Urban and Martin Gerner. Very special thanks go to the German Commission for UNESCO, since without their crucial support in organising and funding the conference, the present volume would not have been possible.

Bibliography

Atapattu, Sumudu, 'Climate Change, Resource Scarcity, Migration and Conflict: Implications for International Peace and Security' forthcoming in id. Human Rights Approaches to Climate Change: Challenges and Opportunities (Routledge, 2015)

–, 'Climate Change: Disappearing States, Migration and Challenges for International Law' Washington Journal of Environmental Law & Policy (forthcoming)

Benjaminsen Tor A, Koffi Alinon, Halvard Buhaug, Jill T Buseth 'Does climate change drive land-use conflicts in the Sahel?' (2012) 49 Journal of Peace Research 97

Brzoska, Michael, Martin Kalinowski, Volker Matthies and Berthold Meyer (eds) *Klimawandel und Konflikte. Versicherheitlichung versus Präventive Friedenspolitik?* (Nomos 2012)

Buhaug, Halvard 'Climate not to blame for African civil wars' (2010) PNAS Early Edition, www.pnas.org/cgi/doi/10.1073/pnas.1005739107 accessed 1 June 2014

Evans, Gareth, Conflict Potential in a World of Climate Change, Address by the President of the International Crisis Group, to Bucerius Summer School on Global Governance 2008, Berlin, 29 August 2008, www.crisisgroup.org/en/publication-type/speeches/2008/conflict-potential-in-a-world-climate-change.aspx accessed 1 June 2014

Fjelde, Hanne and Nina von Uexkull 'Climate triggers: Rainfall anomalies, vulnerability and communal conflict in Sub-Saharan Africa' (2012) 31(7) Political Geography 444

Flint, Julie and Alex de Waal, Darfur: A Short History of a Long War (Zed Books 2005) Gartzke, Erik, 'Could climate change precipitate peace?' (2012) 49(1) Journal of Peace Research 177

Gerner, Martin, *Lead Agency: UNESCO's Global Leadership and Co-ordination Role for the United Nations Decade of Education for Sustainable Development (2005-2014)*, Dresden Papers on Law and Policy of the United Nations Vol 17 (Peter Lang 2013)

Gleditsch, Nils P, Kathryn Furlong, Håvard Hegre, Bethany Lacina & Taylor Owen, 'Conflicts over Shared Rivers: Resource Scarcity or Fuzzy Boundaries?', (2006) 25(4) Political Geography 361

Hirche, Walter, Kulturelle Zusammenarbeit als dritte Dimension der deutschen Außenpolitik, in: Deutsche UNESCO-Kommission (Hrsg.), Lernziel Weltoffenheit, Fünfzig Jahre deutsche Mitarbeit in der UNESCO, 2001

Homer-Dixon, Thomas F., 'On the Threshold: Environmental Changes as a Cause of Acute Conflict' (1991) 16(2) International Security 76

–, 'Environmental Scarcities and Violent Conflict: Evidence from Cases' (1994) 19:1 International Security 5

Isted, Kathryn, 'Sovereignty in the Arctic: An Analysis of Territorial Disputes and Environmental Policy Considerations' (2009) 18(2) Journal of Transnat'l Law & Policy 343

Johnstone, Sarah and Jeffrey Mazo, 'Global Warming and the Arab Spring' (2011) 53(2) Survival: Global Politics and Strategy 11

Kolmannskog, Vikram O, *Future Floods of Refugees: A Comment on Climate Change, Conflict and Forced Migration* (Norwegian Refugee Council 2008)

Lenzerini, Federico, 'Articles 15-16 World Heritage Fund', in Francesco Francioni with Federico Lenzerini (eds.), *The 1972 World Heritage Convention. A Commentary* (Oxford University Press 2008), 269-288

Ludwig, Ralf, Roberto Roson, Christos Zografos and Giorgios Kallis, 'Towards an Interdisciplinary Research Agenda on Climate Change, Water and Security in Southern Europe and Neighboring Countries', (2011) 14 Environmental Science & Policy 794

Maus, Sylvia 'Hand in Hand Against Climate Change: Cultural Human Rights and the Protection of Cultural Heritage' Cambridge Review of International Affairs (forthcoming 2014)

McAdam, Jane, 'Swimming against the Tide: Why a Climate Change Displacement Treaty is Not the Answer' (2011) 23(1) International Journal of Refugee Law 2

Myers, Norman and Jennifer Kent, *Environmental Exodus: An Emergent Crisis in the Global Arena* (The Climate Institute 2005)

O'Loughlina, John, Frank DW Witmer, Andrew M Linke, Arlene Laing, Andrew Gettelman, and Jimy Dudhia, 'Climate Variability and Conflict Risk in East Africa, 1990–2009' (2012) PNAS Early Edition, www.pnas.org/cgi/doi/10.1073/pnas.1205130109 accessed 1 June 2014

Perry, Jim and Chalie Falzon, 'Climate Change Adaptation for Natural World Heritage Sites. A Practical Guide' (2014) UNESCO World Heritage Series No 37

Rayfuse, Rosemary and Emily Crawford, 'Climate Change, Sovereignty and Statehood' (2011) Sydney Law School, Legal Studies Research Paper No. 11/59

Salehyan, Idean, 'From Climate Change to Conflict? No Consensus Yet' (2008) 45(3) Journal of Peace Research 315

Schorlemer, Sabine von, 'Kulturpolitik im Völkerrecht verankert. Das neuen UNESCOÜbereinkommen zum Schutz der kulturellen Vielfalt' (2005) 6 Zeitschrift Vereinte 217

–, 'Cultural Heritage Law: Recent Developments in the Laws of War and Occupation' in James AR Nafziger and Ann M Nicgorski, *Cultural Heritage Issues: The Legacy of Conquest, Colonization and Commerce* (Martinus Nijhoff 2010) 137-158

–, 'and Peter-Tobias Stoll (eds.), *The UNESCO Convention on the Protection and Promotion of the Diversity of Cultural Expressions. Explanatory Notes* (Springer 2012)

Snyder, Robert, International Legal Regimes to Manage Indigenous Rights and Arctic Disputes from Climate Change (2011) 22 Colorado Journal of Int'l Environmental Law & Policy 1

Stern, Nicolas, The Economics of Climate Change (Cambridge University Press 2006)

Sunga, Lyal S, 'Does Climate Change Kill People in Darfur?' (2011) 2(1) Journal of Human Rights and the Environment 64

Wasum-Rainer, Susanne, Ingo Winkelmann and Katrin Tiroch (eds) *Arctic Science, International Law and Climate Change: Legal Aspects of Marine Science in the Arctic Ocean* (Springer 2012)

Werz, Michael and Laura Conley, 'Climate Change, Migration, and Conflict: Addressing Complex Crisis Scenarios in the 21st Century' (Center for American Progress and Heinrich-Böll-Stiftung 2013), www.boell.org/downloads/climate_migration_execsumm.pdf accessed 1 June 2014, 5

Interdisciplinary Approaches of Linking Climate Change with Politics and Law of Cultural Heritage

Bernd von Droste zu Hülshoff*

Man-made Climate Change: A Major Challenge for World Heritage Conservation

No issue is more international than global warming: all people on the planet share the same atmosphere. The world is currently engaged in a grand experiment, studying what happens when you release carbon dioxide into the atmosphere in greater and greater amounts. The scientific community is fairly sure of the result, a reality that is already occurring: glaciers and the polar ice cap will melt, ocean currents will change, and ocean levels will rise. Unlike other problems associated with globalization; however, global environmental problems affect developed and developing countries alike. Climate change will have an impact on social and cultural aspects, with communities changing the way they live, work, worship, and socialise in buildings, sites, and landscapes. Whole communities may be forced to migrate and abandon their built heritage.

I Natural Heritage

Many World Heritage sites already exhibit serious effects due to global warming. In particular, some of the world's most magnificent glaciers – the *Jungfrau, Eiger,* and *Moench* range in Switzerland and the magnificent glacier of *Saint Elias* in Alaska – that are on the World Heritage List. There is also alarming news about the disappearance of corral reefs and their bleaching – for instance at the *Great Barrier Reef* in Australia, the *Moon Reef* in Belize, and the *Galapagos Islands*. Lastly, terrestrial biodiversity, which is the foundation of many natural World Heritage sites, may also be affected through species shifting ranges, changes in the timing of biological cycles, migration of pests, and invasive species.

II Cultural Heritage

In addition to these environmental considerations, there is a growing concern, albeit rather late, over the implications of climate change on the historic built environment. Threats to urban areas and historic cities include direct impacts, such as more frequent and severe flooding due to extreme weather conditions

---

* Former Assistant Director-General of UNESCO and Founding Director of the UNESCO World Heritage Centre.

or rising sea levels. Moreover, a high percentage of the urban population in low-income and middle-income countries live within the Low-Elevation Coastal Zone, the continuous area along the coast that is less than 10 metres above sea level. This creates a danger-zone for such communities.

Ancient buildings were designed for a specific local climate; a climate which is now changing. For instance, climate change is the root cause for desertification. Sand encroachment is increasingly affecting, for example, the famous *Mosques of Timbuktu* in Mali and the *City of Chinguetti* in Mauretania. Furthermore, increasing sea levels threaten many coastal sites, as exemplified by the ever growing number of *Acqua Alta* (high water) days in the World Heritage *City of Venice*. As a final example, increasing soil temperature may degrade the conditions for the conservation of archaeological evidence. Many World Heritage cultural sites have already had to cope with climate change risks and impacts. Some climate change related processes that may have a negative impact on World Heritage sites include: soil moisture such as freeze-thaw cycles, a larger amplitude in temperature and humidity changes, biological infestation of organic building materials, floods, coastal erosion and sea level rise, increased frequency of severe weather, and desertification.

Quantifying these risks and impacts, 72 per cent of State Parties surveyed reported climate change impacts on World Heritage. Furthermore, 125 sites in 59 different countries are allegedly affected.[1] The fact that climate change poses a threat to the outstanding universal values of an increasing number of World Heritage sites has considerable implications for the 1972 World Heritage Convention.[2]

III World Heritage Convention and Committee

Since 2005, the World Heritage Committee has debated the climate change issue from the perspective of Cultural Heritage.[3] The Committee addressed a broad range of basic questions: What is the real scope of the threat? What can the Committee do if changing climatic conditions threaten to destroy the very qualities for which sites were designated World Heritage?

---

1 UNESCO, 'Climate Change and World Heritage' World Heritage Reports 22 (1 May 2007) 26.

2 Convention concerning the Protection of the World Cultural and Natural Heritage; adopted 16 November 1972, entered into force 17 December 1975, 1037 UNTS 151.

3 UNESCO, 'Threats to World Heritage Properties' (9 September 2005) WHC05/29.COM/22, World Heritage Committee Decision 29COM 7B.a.

Most importantly, the World Heritage Committee adopted the "Strategy to assist States Parties to implement appropriate management responses"[4] to assist State Parties in addressing these emerging threats. The main objective of this strategy was to review the main topics that should be considered when preparing to implement preventive and/or corrective management responses to deal with the adverse impacts of climate change. The Committee identified three actions to safeguard heritage from the effects of climate change:

1) *Preventive actions*: monitoring, reporting and mitigation of climate change effects through environmentally sound choices and decisions at a range of levels (individual, community, institutional and corporate).
2) *Corrective actions*: adaptation to the reality of climate change through global and regional strategies and local management plans.
3) *Sharing knowledge*: including best practices, research, communication, public and political support, education and training, capacity building, networking, etc.

The World Heritage Committee climate change strategy also outlined that there are strong links between natural and Cultural Heritage, highlighting that the climate change issue could be used as an opportunity for the two parts of the Convention to be brought closer together. Therefore, whereas climate change impacts will differ for World Heritage of natural and cultural types, the proposed strategy should address both types of properties jointly. The Committee also stressed that climate change is only one risk among a number of challenges facing World Heritage sites. Therefore, this threat should be considered in the broader context of the conservation of these sites. In 2007, the General Assembly of World Heritage States Parties adopted the "Policy Document on the Impacts of Climate Change on World Heritage Properties";[5] which states, among others, that impacts of climate change are already visible; these impacts are better studied in relation to natural sites than cultural places; research should focus on adaptation; and, the iconic value of World Heritage is used for communication and awareness raising. Furthermore, the policy document gave priority to three adaption strategies. There is a need for (1) research and vulnerability assessments; (2) documentation, monitoring and cooperation; and (3) the strengthening of management and continuous maintenance. The policy document also states that in some cases, losses will be unavoidable. Nonetheless, ways to

---

4 UNESCO, 'Strategy to Assist States Parties to Implement Appropriate Management Responses' (26 June 2006) WHC-06/30.COM/7.1.

5 UNESCO, 'Policy Document on the Impacts of Climate Change on World Heritage Properties' (6 June 2008) WHC-07/16.GA/10.

increase resilience include corridors between protected areas, preventive draining of glacial lakes, and maintaining and improving coastal dykes.

The World Heritage Committee also adopted principles guiding research, formulating priorities for research, including priorities for Cultural Heritage.[6] Over the past 35 years the UNESCO World Heritage List has, in a sense, evolved into close to 1000 observation posts around the world. These World Heritage sites, representative of our most outstanding Cultural and Natural Heritage, inevitably register the impact of changes that occur on our planet.

World Heritage sites are continually monitored and assessed. If a site happens to be flooded, the activity is observed, recorded, and the World Heritage Committee is duly informed in order to take appropriate action.

IV UNESCO Climate Initiatives

UNESCO's climate related initiatives are conducted in close collaboration with relevant bodies already involved in climate change and heritage conservation issues, such as the United Nations Framework Convention on Climate Change (UNFCCC)[7], the Intergovernmental Panel on Climate Change (IPCC)[8], the Convention on Biological Diversity (CBD)[9], the UNESCO Man and the Biosphere Programme (MAB)[10] , the Ramsar Convention on Wetlands[11], and the UNESCO conventions dealing with Cultural Heritage.[12]

In addition, the UNESCO World Heritage Centre has undertaken a number of activities relating to climate change issues. One of the most important is the

6 The UNESCO World Heritage Centre issued a number of World Heritage and Climate Change Publications which can be consulted at whc.unesco.org/en/climate change accessed 25 September 2013.

7 Adopted 9 May 1992, entered into force 21 March 1994, 1771 UNTS 107 (2013: 195 State Parties).

8 Created in 1988; see www.ipcc.ch accessed 25 September 2013.

9 Adopted 5 June 1992, entered into force 29 December 1993, 1760 UNTS 79 (2013: 193 State Parties).

10 Created in 1971; see www.unesco.org/new/en/natural-sciences/environment/ecological sciences/man-and-biosphere-programme accessed 25 September 2013.

11 Convention on Wetlands of International Importance especially as Waterfowl Habitat; adopted 2 February 1971, entered into force on 21 December 1975, 996 UNTS 245 (2013: 168 State Parties).

12 For an overview see UNESCO, 'Protecting Our Heritage and Fostering Creativity' en.unesco.org/themes/protecting-our-heritage-and-fostering-creativity accessed 25 September 2013.

report "Case studies on climate change and World Heritage".[13] This report features 26 sites, including the Tower of London, Kilimanjaro National Park, and the Great Barrier Reef. These case studies are representative of some of the dangers faced by the nearly 1000 properties inscribed on the World Heritage List.

UNESCO also addresses the climate change issue in general. In 2009, the UNESCO Director-General, Irina Bokova, launched the UNESCO Climate Change Initiative.[14] It seeks to reinforce the scientific, mitigation, and adaptation capacities of countries and communities that are most vulnerable to the effects of climate change.

V Conclusion

Let me further underline that rich countries have a particular obligation to commit to protecting our common global Cultural Heritage. It is the poor countries that bear the greatest risks related to climate change: on the one hand for geographical reasons and on the other hand because they are less protected against destruction and diseases caused by climatic catastrophes. Furthermore, many of the poor have no insurance to pay for damage and remedies. Without the help of developed countries, they will not be able to conserve World Heritage as effectively, if at all.

We hold the natural and cultural environment of our planet in common, both with other members of the present generation and with other generations, past and future:

> "At any given time, each generation is both custodian and trustee of the planet for future generations and a beneficiary of its fruits. This imposes obligations upon us to care for the planet, but also gives us certain rights to use it".[15]

Planetary obligations such as caring for the world climate and caring for our cultural and natural inheritance are based on the principles of equity between generations. These obligations become enforceable when they are made specific by international agreements, national and local laws, transformed into customary international law, or adopted as general principles of law. They must ultimately be defined and applied in the context of specific problems such as conserving World Heritage in a time of accelerated climate change. Let us assume this important task.

---

13 UNESCO, 'Case Studies on Climate Change and World Heritage' (UNESCO World Heritage Center 2007) whc.unesco.org/en/activities/473 accessed 25 September 2013.

14 www.unesco.org/new/fileadmin/MULTIMEDIA/HQ/SC/pdf/sc_climChange_initiative_EN.pdf accessed 25 September 2013.

15 Edith Brown Weiss, *In Fairness to Future Generations: International Law, Common Patrimony, and Intergenerational Equity* (Transnational Publishers 1989) 17.

Bibliography

UNESCO, 'Case Studies on Climate Change and World Heritage' (UNESCO World Heritage Center 2007) whc.unesco.org/en/activities/473 accessed 25 September 2013

–, 'Climate Change and World Heritage' World Heritage Reports 22 (1 May 2007)

–, 'Policy Document on the Impacts of Climate Change on World Heritage Properties' (6 June 2008) WHC-07/16.GA/10

–, 'Strategy to Assist States Parties to Implement Appropriate Management Responses' (26 June 2006) WHC-06/30.COM/7.1

Weiss, E B, *In Fairness to Future Generations: International Law, Common Patrimony, and Intergenerational Equity* (Transnational Publishers 1989)

Gunter Pleuger*

Climate Change as a Threat to International Peace – The Role of the UN Security Council

Climate Change has been recognized as a global challenge that needs action of the international community of states. The Kyoto-Protocol,[1] the Millennium Declaration,[2] treaties and international agreements, resolutions of the General Assembly of the United Nations and regional efforts have been made to address the problem – but so far with no comprehensive and sustainable effect.

One of the main obstacles is the lack of legal and enforceable obligations of states, leaving the actual implementation of agreements to the sovereign decision of governments.

Final documents of conferences and international or regional arrangements are normally politically binding only to those states who have signed the documents, and these normally do not contain the threat of sanctions in case of noncompliance. Therefore, the implementation of climate goals varies between efforts and neglect.

Climate change as a global problem needs global solutions; and global solutions need a global institution that takes the necessary decisions. The only global institution that we have is the United Nations. However, the resolutions of the General Assembly are not legally, but only politically, binding on those member states who have voted *Yes*. Those who have voted *No*, or abstained, have no political obligation to implement the resolution adopted by the majority of member states in the General Assembly.

The only body of the United Nations that can make decisions legally binding on all member states is the Security Council. According to Article 25 UN Charter "(t)he Members of the United Nations agree to accept and carry out the decisions of the Security Council in accordance with the present Charter".[3]

The unique decision making position of the Security Council is underlined by Article 12 UN Charter, preventing the General Assembly from making decisions on a subject once the Security Council is dealing with the matter.

---

* President of the European University Viadrina Frankfurt (Oder) and former German Ambassador to the United Nations.

1 Adopted 11 December 1997, entered into force 16 February 2005, 2303 UNTS 148.

2 Adopted and entered into force 08 September 2000, A/RES/55/2.

3 Adopted 26 June 1945, entered into force 24 October 1945, 1 UNTS XVI.

There are two possibilities of action for the Security Council: under Chapter 6 of the Charter it can make recommendations for the pacific settlement of disputes that endanger the maintenance of international peace and security. If that does not help, the Security Council can impose economic or political sanctions, and if necessary, authorize military action. However, Article 39 requires that the Security Council first determine the existence of a threat or breach of peace before taking action under Chapter 7. This gives the Security Council a legally binding decision making power, which is only *ad hoc* in the case of an actual threat to peace and security.

In the past decades, however, the Security Council has embarked on a new policy. It has taken up political and humanitarian issues such as "Children in Armed Conflicts"[4] and has set general rules for actual problems, thereby creating new international law and establishing itself as a new source of international law next to international customary and treaty law.

The foundations for this development were laid by two resolutions: the anti-terrorism resolution[5] and the resolution against the proliferation of weapons of mass-destruction.[6] Both resolutions contain strict rules to be obeyed by all member states of the United Nations and include – where necessary – the obligation to change national legislation. The implementation of the rules set by these resolutions is supervised by two subcommittees of the Security Council. Member states have to report to the Security Council which in the case of noncompliance of a member state, can either order assistance in the case of lack of capability or adopt sanctions in the case of lack of political willingness to implement Security Council resolutions.

This newly developed capacity of the Security Council to create international law will certainly be further developed and used in the future in order to cope with imminent and serious global problems. For the fight against terrorism or against the spread of weapons of mass destruction, and similar threats, it would take too long to create the relevant customary or treaty law.

It would therefore be possible and potentially effective to make use of the powers of the Security Council to set new legal rules in order to cope with climate change. There are, however, a few obstacles.

---

4 UNSC Res 2068 (2012) UN Doc S/RES/2068; UNSC Res 1998 (2011) UN Doc S/RES/1998; UNSC Res 1882 (2009) UN Doc S/RES/1882; see also UNGA 'The Rights of the Child' Res 51/77 (20 February 1997) UN Doc A/RES/51/77.

5 UNSC Res 1373 (2001) UN Doc S/RES/1373.

6 UNSC Res 1540 (2004) UN Doc S/RES/1540.

First the issue of climate change as a threat to peace would have to be set on the Security Council agenda. Traditionally, the agenda of the Security Council is decided by consensus. As the setting of the agenda is a procedural matter, with no right of veto, the five permanent members have made sure that they cannot be outvoted. But, consensus gives every Security Council member a veto.

In a case of widely differing interests among the Security Council members, it can therefore prove difficult to find consensus.

In the Darfur conflict, for instance, two permanent Security Council members with special interests in Sudan prevented the inclusion of Darfur in the Security Council agenda, which in the end was achieved only by a trick of the Security Council presidency proposing as agenda item a "Report on the Situation of Human Rights in Africa" by the UN High Commissioner for Human Rights.

This incident also reflects another obstacle: the fact that the five permanent members with their right of veto dominate the Security Council by their status and often put their national interests above the interests of the international community.

The Security Council therefore, needs the reform that has been discussed for more than 20 years in order to enhance the legitimacy and the effectiveness of its decisions.

At present it might be difficult to engage the Security Council in deliberation and action with regard to climate change as a threat to peace, but it is worth to try and seek a coalition in the United Nations to put this issue on the Security Council agenda. That could open new and creative ways, and effective rules, to resolve this global challenge.

Franziska Knur*

The United Nations Human Rights-Based Approach to Climate Change – Introducing a Human Dimension to International Climate Law

Abstract The article revisits the 2009 Office of the High Commissioner for Human Rights Report on the relationship of climate change and human rights and undertakes to develop the United Nations human rights-based approach to climate change into a more meaningful concept. Accepting the negative impact of global warming on the enjoyment of human rights States could be obliged to reduce green house gas emissions and provide substantial international assistance to the States most affected by the consequences of global warming, not merely based on the international climate regime, but on the basis of international human rights law. Human rights also provide an accountability framework for damages related to climate change once the conceptual link of human rights law and global warming is accepted. Although approaches to enhance the concept remain fragmented, it is argued that a human rights-based approach can have a significant impact on the international climate change discourse: It provides access to institutions for the concretization of State obligations and introduces subjective rights and thus a vertical dimension to international climate law.

I Introduction

Appropriately addressing climate change is one of today's major challenges to the international community since strategies to limit human-induced global warming as well as to adapt to the consequences of global climate change require international co-operation. Thus, the international climate regime presents itself as a

* Doctoral candidate and research assistant at the Chair for Public Law, European and International Law at Dresden University of Technology.
This article is based on a presentation at the Junior Research Panel convened by the Centre for International Studies of the TU Dresden on 1 June 2013. I would like to thank Constanze Zahm for her valuable comments during her rejoinder at the symposium.

"highly specialized area of international law".[1] The 1992 United Nations Framework Convention on Climate Change (UNFCCC)[2], complemented by the 1997 Kyoto-Protocol[3], acknowledges "that change in the Earth's climate and its adverse effects are a common concern of humankind" (1st preamble clause) and provides for a common international legal framework in order to stabilize "greenhouse gas concentrations in the atmosphere at a level that would prevent dangerous anthropogenic interference with the climate system" (Article 2 UNFCCC). In addition to these uniquely formulated treaty provisions, preexisting rules of international law – e.g. rules of international environmental or human rights law as well as secondary rules with regard to State responsibility – frame the legal context of international climate change regulation.[4] The identification of pertinent international rules as well as the determination of the scope of their applicability in the climate change context requires a comprehensive interpretation process governed by a variety of actors such as States, international organizations, non-governmental organizations (NGOs), international jurisprudence and legal scholars.

This article elaborates on the undertaking of the United Nations, especially its Office of the High Commissioner for Human Rights (OHCHR), in order to implement a human rights-based approach to climate change. The conceptional link of human rights and climate change goes back to an initiative of the Maldives.[5] In November 2007 a group of Small Island Developing States adopted the "Male Declaration on the Human Dimension of Global Climate Change"[6] requesting the assessment of human rights implications of global warming. The Human Rights Council adopted this approach in 2008 and invited OHCHR to draft a study on the issue.[7] Based on the written and oral submissions of States, international institutions, NGOs and scientific groups, the OHCHR's "Report

---

1 See e.g. Erkki J Hollo, Kati Kulovesi and Michael Mehling, 'Climate Change and the Law' in Erkki J Hollo, Kati Kulovesi and Michael Mehling (eds), *Climate Change and the Law* (Springer 2012) 2.

2 Adopted 9 May 1992, entered into force 21 March 1994, 1771 UNTS 107.

3 Adopted 11 December 1997, entered into force 16 February 2005, (1998) 37 ILM 22.

4 See e.g. Siobhan McInerney-Lankford, Mac Darrow and Lavanya Rajamani, *Human Rights and Climate Change: A Review of the International Legal Dimensions* (World Bank Publications 2011) 11–19.

5 On the role of the Maldives see John Knox, 'Linking Human Rights and Climate Change at the United Nations' (2009) 33 Harv Envt'l L Rev 477, 479–484.

6 (13–14 November 2007) www.ciel.org/Publications/Male_Declaration_Nov07.pdf accessed 18 June 2013.

7 UNHRC Res 7/23 (28 March 2008) UN Doc A/HRC/7/78.

on the relationship between climate change and human rights" was published in January 2009.[8] It was presented before the Human Rights Council, which took note of the study in resolution 10/4 of 25 March 2009 and which decided to hold an interactive panel discussion on the relationship between climate change and human rights.[9] At the debate held in June 2009 the Council concluded that it was "valuable to look at climate change-related effects from a human rights perspective".[10] It became, however, evident that there was substantial disagreement between States on the question whether – and if, to what extent – international human rights law contains State obligations in the context of addressing the causes and consequences of climate change. It remained unanswered how the normative construction of correlating rights and duties of international human rights law could be linked in a meaningful way to the international effort to manage climate change.

This article reopens the debate and discusses the advantage of the United Nations human rights-based approach to climate change. First, the key findings of the OHCHR report will be revisited. With reference to three key questions, secondly it will be assessed how the human rights-based approach could further be developed into a more meaningful concept. Third, it will be suggested that the human rights-based approach uniquely contributes to the international discourse on climate change by introducing a human dimension to international climate law. On a practical as well as on a conceptual level, it will be argued that the acceptance of human rights as a cross-cutting issue within the context of global warming will – in the long run – be characteristic for the further development of the international system.

II The Report of the OHCHR on the Relationship between Climate Change and Human Rights

The OHCHR report discusses "how the observed and projected impacts of climate change have implications for the enjoyment of human rights and for the obligations of States under international human rights law".[11] However, it does

---

8 (15 January 2009) UN Doc A/HRC/10/61.

9 UNHRC Res 10/4 (25 March 2009) UN Doc A/HRC/10/29.

10 UNHRC 'Panel Discussion on the Relationship between Climate Change and Human Rights: Summary of Discussions' (15 June 2009) www.ohchr.org/Documents/Issues/ClimateChange/SummaryPanelDiscussion.doc accessed 7 June 2013.

11 (15 January 2009) UN Doc A/HRC/10/61 Summary.

not provide a definite answer to the most obvious question, that is, whether climate change in itself constitutes a human rights violation.[12] The report states:

> "While climate change has obvious implications for the enjoyment of human rights, it is less obvious whether, and to what extent, such effects can be qualified as human rights violations".[13]

In answering this cardinal question, a twofold approach can be followed since international human rights law has an *entitling* as well as an *obligating* dimension.[14]

1. The Entitling Dimension of Human Rights Law and Climate Change

First of all, the international system for the promotion and protection of human rights departs from the idea that all human beings are *entitled* to the rights and freedoms proclaimed by the Universal Declaration of Human Rights[15] and contained in international legal instruments such as the International Covenant on Civil and Political Rights (ICCPR)[16] and the International Covenant on Economic, Social and Cultural Rights (ICESCR).[17]

This entitling dimension of human rights law, with regard to climate change, is summarized by the OHCHR report under the heading "implications for the enjoyment of human rights". It states that global warming will "potentially have implications for the full range of human rights" (Paragraph 20) and further mentions to that extent, explicitly, the right to life, right to adequate food, right to health and right to housing as well as the right to self-determination. Additionally, it emphasizes that the effects of climate change will be felt most acutely by vulnerable groups such as women, children and indigenous peoples and that these groups are entitled to special protection in accordance with the principle of equality and non-discrimination (Paragraph 42).

---

12 Likewise Clemens Müller and Kristine Franzen, 'Der Klimawandel und das Menschenrechtssystem der Vereinten Nationen' (2010) 4 Zeitschrift für Menschenrechte 7, 14.

13 (15 January 2009) UN Doc A/HRC/10/61 Paragraph 70.

14 Walter Kälin and Jörg Künzli, *Universeller Menschenrechtsschutz* (2nd ed, Helbing-Lichtenhahn 2008) 89.

15 Adopted 10 December 1948, UNGA Res 217 A (III).

16 Adopted 16 December 1966, entered into force 23 March 1976, 999 UNTS 171.

17 Adopted 16 December 1966, entered into force 3 January 1976, 993 UNTS 3. See also the wording of the rights contained in the ICCPR, e.g. "Every human being has the inherent right to life" (Article 6), "Everyone has the right to liberty and security of person" (Article 9 Paragraph 1).

Overall, it is not controversial that climate change does and continues to affect the enjoyment of human rights.[18] Thus, from the perspective of the ones entitled by human rights, it can be concluded that climate change and its consequences violate human rights.[19]

2. The Obligating Dimension of Human Rights Law and Climate Change

A different question is whether climate change can be qualified as a human rights violation that is legally attributable to States as the primary duty-bearers under international human rights law. After all, a key feature of the international human rights system is the creation of State *obligations*.[20] A State owes obedience to human rights provisions contained in international covenants and customary law towards the individual person, bilaterally towards another State and at all times to the whole international community (*erga omnes/erga omnes partes*).[21] If States violate their human rights obligations, secondary norms of international law become applicable.[22] Therefore, it stands to question whether contributing to climate change by emitting green house gases – and thus contributing to the harmful consequences of global warming – constitutes a violation of States human rights obligations.

The OHCHR identifies three obstacles to this argumentation: First, it would be "impossible to disentangle the complex causal relationships linking historical greenhouse gas emissions of a particular country with a specific climate change-related effect, let alone with the range of direct and indirect implications for human rights"; second, global warming would often be only "one of several contributing factors to climate change-related effects"; and third, harmful consequences of global warming would often be "projections about future impacts, whereas human rights violations are normally established after the harm has occurred" (Paragraph 70).

---

18 See the statements of State representatives contained in UNHRC, 'Summary of Panel Discussion' (n 10) Paragraphs 33–54.

19 For a 'survey of illustrative examples' see McInerney-Lankford, Darrow and Rajamani (n 4) 11–19.

20 See again the wording of the international covenants such as "Each State Party undertakes to respect and to ensure (…)" (Article 2 Paragraph 1 ICCPR); see also Kälin and Künzli (n 14) 90–94.

21 See Kälin and Künzli (n 14) 107–109.

22 However, see on human rights as self-contained regimes Eckart Klein, 'Self-Contained Regime' in Rüdiger Wolfrum (ed), *The Max Planck Encyclopedia of Public International Law* (2008 online ed, accessed 6 July 2013) Paragraph 14.

Before turning to a discussion of these arguments and the assessment of conceptual approaches to further develop the relationship of human rights and climate change, it will be briefly revisited where the OHCHR sees the *value added* through a human rights-based approach to climate change: International human rights law provides States with normative guidance in protecting individuals and adopting climate change response measures (Paragraph 71). Discussing the relevant human rights obligations of States, the report distinguishes between national and international human rights obligations.

a) *National Level Obligations of States Dealing with Climate Change*

The report emphasizes the national level obligations deriving from international human rights law. Foremost, States are obliged to protect their inhabitants from foreseeable threats related to climate change (Paragraph 74). Also, when adopting measures in response to climate change, States need to act in accordance with their domestic human rights obligations. Thus, mitigation measures shall take into account possible implications on the enjoyment of human rights. Likewise, adaptation measures need to be human rights-compliant and substantial procedural safeguards, such as the right to access to information and the right to participation in decision making processes, shall be ensured (Paragraphs 78–79).

b) *International Level Obligations of States in the Context of Climate Change*

Additionally, the report derives some international level State obligations in the context of climate change from human rights law. Especially from an international equity perspective, it is desirable not only to impose obligations to deal with the consequences of global warming in a human rights-compliant way on the countries most affected by climate change, but also to remind industrialized countries of their responsibilities under international human rights law.[23] Of course, the discussion within the Human Rights Council on this issue proved difficult. Whereas developing countries and Small Island States did not want to be 'left alone', industrialized countries insisted on emphasizing on national level obligations.[24]

---

23 See the statements of State representatives on international assistance and cooperation as well as on equity under the UNFCCC and human rights contained in UNHRC, 'Summary of Panel Discussion' (n 10) Paragraphs 68–92.

24 See on this divide Marc Limon, 'Human Rights Obligations and Accountability in the Face of Climate Change' (2010) 38 Ga J Int'l Comp L 543, 550; see as an example for

However, according to OHCHR there undoubtedly exists an obligation for international co-operation and assistance in international law that also applies to the context of climate change (Paragraphs 84–85). This international obligation might be derived from Article 1 Paragraph 3 United Nations Charter as well as from Article 2 ICESCR. Therefore, States are – also due to their knowledge on the human rights implications of global warming – obliged to work together and to "take steps through international assistance and co-operation, depending on the availability of resources, to facilitate fulfilment of human rights in other countries" (Paragraph 86).

3. Discussion

The report, although generally welcomed, has been the object of criticism.[25] While drawing a thorough picture of how climate change may affect the enjoyment of human rights, its elaborations on State obligations fall rather short. Although it prominently features the fact that climate change mitigation and adaption strategies contain human rights violating potential, national as well as international State obligations should have been elaborated in more detail.[26] Additionally, the determination of extra-territorial State obligations should have been included.[27] Thus, the report underlines the apprehension that human rights have become the "common currency of contemporary legal and political discourse";[28] at the same time substantial room for further elaborations remains in order to turn the human rights-based approach into a meaningful concept.

III Three Conceptual Questions for the Further Development of the Human Rights-Based Approach

In order to investigate how the human rights-based approach to climate change might reach its full potential, three conceptual questions will be elaborated upon:

---

the reluctance of industrialized States John von Doussa, Allison Corkey and Renée Chartres, 'Human Rights and Climate Change' (2007) 14 Austl Int'l L J 161, 174–176.

25 See e.g. Limon, 'Human Rights Obligations and Accountability' (n 24) 586; Knox, 'Linking Human Rights and Climate Change' (n 5) 496.

26 Ibid 478.

27 John Knox, 'Climate Change and Human Rights Law' (2009) 50 Va J Int'l L 163, 200; Müller and Franzen (n 12) 17.

28 Dimitris Efthymiou, 'Climate Change, Human Rights and Distributive Justice' (2009) 20 Finnish YBIL 111, 112.

First, it will be asked whether a State duty to reduce green house gas emissions could be derived from human rights law. Second, it will be considered how States could be held responsible *id est* liable for the harmful consequences of human-induced global warming on the grounds of human rights law. Third, the human rights-based duty to provide international assistance will be revisited.

1. Human Rights-Based Duty to Reduce Green House Gas Emissions

With the negative effects of hand-made climate change on the enjoyment of human rights in mind, one might ask whether States are obliged to reduce green house gas emissions on the basis of their human rights obligations.[29] Since it has become evident that emissions contribute to a human-induced process of global warming, how are States allowed to do business as usual with regard to their national emission policies? From a human rights law perspective, can they openly accept the obvious impacts of their emissions on the enjoyment of human rights around the globe? At least under the ICESCR, States are obligated to promote human rights internationally.[30] Every action that obviously hinders higher standards of living should thus be regarded irreconcilable with international human rights law. Therefore, a rule to abstain from policies that allow further contribution to global warming should be acknowledged. Of course, it is difficult to derive detailed emission reduction goals from human rights provisions. This is not to be undertaken here; rather, attention is drawn to the fact that it is within the competences of human rights institutions to explore this issue. There is a variety of human rights institutions that can, and should, deal with the question of how States' human rights obligations are linked to international climate law and their obligation to reduce greenhouse gas emissions.[31]

29 Arguing strongly in favor of such a duty: Food First Information and Action Network Deutschland, 'Menschenrechte im Klimawandel – Anforderungen an die deutsche und internationale Klimapolitik 2009' www.fian-deutschland.de/online/index.php?option=com_remository&Itemid=160&func=startdown&id=16 5 accessed 7 June 2013.

30 According to Article 2 ICESCR States "undertake steps with a view to achieving the realization of the rights" in general. Thus, the provisions are at least to some extent extraterritorially applicable; see for an examination of the extraterritorial State obligations of the ICCPR and the ICESCR with regard to climate change Knox, 'Climate Change and Human Rights Law' (n 27) 202–206.

31 See for the competences of the Human Rights Council UNGA Res 60/251 (3 April 2006) UN Doc A/RES/60/251 Paragraphs 2–6; for the evolutionary interpretation of the competences of the Human Rights Committee and the ICESCR Helen Keller and Leena Grover, 'General Comments of the Human Rights Committee and Their

On the one hand, the UN-charter-based human rights institutions should renew their commitment to investigating the topic of climate change and human rights. As has previously been shown, the debates on climate change and human rights at the Human Rights Council are highly controversial – here, the restrictions of political human rights bodies have become evident.[32] However, (further) discussion within this forum keeps the issue on the international agenda.[33] When States publicly argue that it is not their obligation from human rights law to reduce emissions, it gives NGOs and international civil society the chance to disagree in public.[34] Again, the OHCHR may act as a think tank providing not only the members of the Human Rights Council with facts and figures but also suggesting further legal evaluations and insights. It might also be advisable to open a thematic procedure. By appointing a special rapporteur, an expert would further a "comparative and global understanding of the human rights problem"[35] and suggest how single human rights obligations could extent to a duty to reduce emissions.[36]

On the other hand, all major international human rights conventions are equipped with monitoring institutions consisting of international experts.[37] Each convention committee is competent to evaluate how climate change affects the enjoyment of the relevant convention rights, currently and in the near future. By establishing this link, it is to be elaborated to what extent the pertinent conventions contain State obligations that might also affect national climate and emission policies. Here, a valuable tool "to give states guidance in nature and scope of other

---

Legitimacy' in Helen Keller and Geir Ulfstein (eds), *UN Human Rights Treaty Bodies: Law and Legitimacy* (Cambridge University Press 2012) 127.

32 See n 24.

33 See Nigel S Rodely, 'UN Treaty Bodies and the Human Rights Council' in Helen Keller and Geir Ulfstein (eds), *UN Human Rights Treaty Bodies: Law and Legitimacy* (Cambridge University Press 2012) 352.

34 See e.g. the report of the Food First Information and Action Network (n 29).

35 Rodely (n 33) 353.

36 See also the work of the thematic rapporteurs assigned by the Human Rights Council (and its predecessor, the Human Rights Commission) already conducted on climate change as reviewed in Müller and Franzen (n 12) 17–23.

37 Such as the Human Rights Committee (Article 28 ICCPR), Committee on Economic, Social and Cultural Rights (ECOSOC Res 1985/17 (28 May 1985) UN Doc E/RES/1985/17), Committee on the Elimination of Discrimination against Women (Article 17 Convention on the Elimination of all Forms of Discrimination against Women), Committee on the Rights of the Child (Article 43 Convention on the Rights of the Child) etc.

obligations"[38] are general comments that could address the relationship between pertinent human rights and the global threat of climate change. Although not formally legally binding, general comments have a robust legal analytical function by defining the scope of convention rights.[39] Furthermore, they often contain policy recommendations[40] and thus might suggest to what extent States should cut emissions and what measures are advisable to take in order to reverse the negative effects of national emission policies on the human rights situation around the globe.

2. International Responsibility and Liability for Harmful Consequences of Human-Induced Global Warming on the Basis of International Human Rights Law

A second question derived from a human rights perspective on climate change is whether States could be held responsible, *id est* liable individually or collectively, for the harmful consequences of human-induced global warming. Generally, the issue of climate change litigation and compensation claims of individuals has not yet been fully explored.[41] International law does not *per se* negate individual claims against States; the Articles on State Responsibility for Internationally Wrongful Acts by the International Law Commission leave this question open (Article 33 Paragraph 2).[42] However, individuals are not procedurally equipped with many opportunities besides seeking diplomatic protection.[43] Also, the UNFCCC remains more or less silent on the issue of responsibility for damage and compensation claims for injury.[44]

---

38 Rodely (n 33) 353.

39 See Keller and Grover (n 31) 129–133.

40 Of course the means and methods of domestic implementation of the covenant rights are to the discretion of the parties (Article 2 ICCPR); see on this Keller and Grover (n 31) 124–125.

41 See Timo Koivurova, 'International Legal Avenues to Address the Plight of Victims of Climate Change: Problems and Prospects' (2007) 22 J Envt'l L & Litig 267, 278–282; see also Christina Voigt, 'State Responsibility for Climate Change Damages' (2008) 77 NJIL 1, 2.

42 UN GAOR (53rd Session 2001) UN Doc A/CN.4/L.602/Rev.1; on the few opportunities for advancing climate change litigation in general international law see Koivurova (n 40) 278–282.

43 Roda Verheyen, *Climate Change Damage and International Law: Prevention Duties and State Responsibility* (Nijhoff 2005) 242.

44 See Hannah Stallard, 'Turning up the Heat on Tuvalu: An Assessment of Potential Compensation for Climate Change Damage in Accordance with States Responsibility under International Law' (2009) 15 Canterbury L Rev 163, 178.

Thus, in addition to domestic lawsuits,[45] human rights law could provide an accountability framework for damages in relation to climate change.[46] This was put to a test when, in 2005, representatives of US-American and Canadian Inuit petitioned to the Inter-American Commission on Human Rights to hold the United States responsible for the violation of human rights.[47] It was argued that the US was responsible for human induced global warming since it is the world's largest emitter of green house gases.[48] Although acknowledging the human causation of climate change, the US-Government would not reduce emissions and denied participation in the Kyoto-Protocol. Since as a result of global warming the Inuit's fundamental rights, e.g. to use and enjoy their traditional lands, to enjoy their personal property, to health and to life, were violated, the claimants requested a declaration that the US be internationally responsible for the violation of these rights; that it adopts measures to limit its greenhouse gas emissions and that it implements a plan to protect Inuit culture and resources.[49]

The petition was found to be inadmissible before the Commission.[50] According to its decision, it was not possible to relate harmful emissions to one single State and to further calculate the national responsibility for climate change.[51] However, the case opened significant debate on the question whether States are responsible *id est* liable for infringements of human rights caused by climate change.[52] Whereas the question of causation still remains unanswered,

---

45 See Michael G Faure and André Nollkaemper, 'International Liability as an Instrument to Prevent and Compensate for Climate Change' (2007) 26 Stan Envt'l L J 123, 147–150; further examples in McInerney-Lankford, Darrow and Rajamani (n 4) 58.

46 See the Statement of Mauritius at the Panel on Human Rights and Climate Change (HRC 11th Session, 15 June 2009) as cited in Limon, 'Human Rights Obligations and Accountability' (n 24) 578.

47 Innuit Circumpolar Council, 'Petition to the Inter-American Commission on Human Rights Seeking Relief from the Violations Resulting from Global Climate Warming Caused by Acts and Omissions of the United States' (7 December 2005) www.inuitcircumpolar.com/files/uploads/icc-files/FINALPetitionSummary.pdf accessed 6 June 2013.

48 Ibid 6.

49 Ibid 4–8.

50 See Jessica Gordon, 'Inter-American Commission on Human Rights to Hold Hearing after Rejecting Inuit Climate Change Petition' (2007) 7 Sustainable Dev L & Pol'y 55.

51 For further discussion of the case see Faure and Nollkaemper (n 45) 157.

52 See e.g. James D Ford, 'Supporting Adaptation: A Priority for Action on Climate Change for Canadian Inuit' (2008) 8 Sustainable Dev & Pol'y 25; Katherine King, 'Climate Change and the Inuit: A Melting of Actions into a Cloudy Mess' (2009) 17 Se Envt'l L J 481.

there is the idea to attribute climate change-related damage to a State based on its historical emissions or to the degree in which it fulfils its reduction targets contained in the Kyoto-Protocol.[53] Some authors also introduce a whole system of reparations based on those targets.[54] However, there is one important obstacle: If the attribution of responsibility for damages is linked to the reduction targets that States willingly consent to, then there is considerably less incentive to commit to such obligations. Also, one could argue that there is not much value added from blaming, rather, the international community should focus on developing common approaches in order to manage future challenges of global warming.

With all these arguments in mind, human rights institutions nevertheless should elaborate on the principles for the international responsibility and liability of States with regard to the effects that climate change has on the enjoyment of human rights.[55] It is imperative to the human rights idea not only to speak of an entitlement to human rights but also to hold States responsible where they do not live up to their obligations. States should at least be evaluated with regard to how they meet protection obligations and adopt appropriate response measures. Thus, assessing individual liability claims for climate change-related damages, not the amount of emissions would be the appropriate yardstick but a State's performance with regard to fulfilling its national level as well as its international level human rights obligations in the context of global warming.[56] On a case to case basis, the extent of State responsibility has to be elaborated by the pertinent judicial and quasi-judicial human rights bodies.[57] Although there are significant procedural obstacles, filing suits and submitting petitions before international and national bodies would initiate further exploration of the possibilities of international climate change litigation.[58] As witnessed in the Inuit case, this

53 See Limon, 'Human Rights Obligations and Accountability' (n 24) 571.

54 Maxine Burkett, 'Climate Reparations' (2009) 10 Melb J Int'l L 509, 521–529.

55 See e.g. for the concept of joint and several liability Faure and Nollkaemper (n 45) 165.

56 See Knox, 'Climate Change and Human Rights Law' (n 27) 209–210.

57 The relevance and impact of decisions of human rights institutions has also been recognized by the ICJ in *Case Concerning Ahmadou Sadio Diallo* (Republic of Guinea v Democratic Republic of Congo) (Judgment of 30 November 2010) (2010) ICJ Rep 639, Paragraph 66.

58 For a critical discussion of the assumption, that human rights litigation may drive global green house gas policy see Eric A Posner, 'Climate Change and International Human Rights Litigation: A Critical Appraisal' (2007) 155 U Pa L Rev 1925.

provides for international publicity. Further, it becomes an avenue for political pressure and a place for NGOs and civil society to present their points of view.[59]

In addition to regional human rights bodies, especially bodies with an international scope seem to be the appropriate place for those kinds of procedures.[60] Complementing the individual complaint procedure before the Human Rights Committee, the recently established mechanism under the ICESCR will provide for valuable institutional resources since social and economic rights in particular are impaired by the consequences of global warming.[61]

3. Human Rights-Based Duty to Provide International Assistance to Affected States

The third question is whether States have any further duties to provide international assistance on the basis of their human rights obligations. Especially, it stands to argue that industrialized countries have a substantial extra-territorial duty to assist the affected countries in the process of adapting to the consequences of climate change. The duty to assist with adaptation (e.g. by means of technology transfer) is also contained in the UNFCCC (Article 4 lit c). In the past, however, it has proven difficult to bring these words to life.[62] Human rights law could help concretize this obligation since it also contains the principle of international cooperation and mutual assistance.[63] Article 2 Paragraph 1 ICESCR reads:

---

59 See Philippe Cullet, 'Liability and Redress for Human-Induced Global Warming: Towards an International Regime' (2007) 43 Stan J Int'l L 99; see also Faure and Nollkaemper (n 45) 179.

60 See Geir Ulfstein, 'Individual Complaints' in Helen Keller and Geir Ulfstein (eds), *UN Human Rights Treaty Bodies: Law and Legitimacy* (Cambridge University Press 2012) 92–100.

61 Optional Protocol to the International Covenant on Economic, Social and Cultural Rights; adopted 10 December 2008, entered into force 5 May 2013, UN Doc A/63/435; see on the new mechanisms Rainer Grote, 'The Optional Protocol to the International Covenant on Social, Economic and Cultural Rights: Towards a More Effective Implementation of Social Rights?' in Holger P Hestermeyer et al (eds), *Coexistence, Cooperation and Solidarity, Liber Amicorum Rüdiger Wolfrum* (Vol I, Nijhoff 2012); on the justiciability of economic, social and cultural rights see Urfan Khaliq and Robin Churchill, 'The Protection of Economic and Social Rights: A Particular Challenge?' in Helen Keller and Geir Ulfstein (eds), *UN Human Rights Treaty Bodies: Law and Legitimacy* (Cambridge University Press 2012) 200–202.

62 See McInerney-Lankford, Darrow and Rajamani (n 4) 61–62.

63 Ibid.

> Each State Party to the present Covenant undertakes to take steps, individually and through international assistance and co-operation, especially economic and technical, to the maximum of its available resources, with a view to achieving progressively the full realization of the rights recognized in the present Covenant by all appropriate means, including particularly the adoption of legislative measures.

Thus, the ICESCR obligation to promote social, economic and cultural human rights is not restricted to the territory of a State party, but extends to international assistance and co-operation. However, it is unclear what is contained in the human rights-based duty to co-operate; the nature of the concept and its legal value is still debated.[64] In terms of providing economic or technical assistance from one State to another, the duty to assist does not emerge until the sovereign State asks for it; even then it has to be realized step by step only.[65] However, the duty of international co-operation and assistance could also contain a 'standing' State obligation to offer assistance to those States and peoples most affected by the consequences of global warming. Should not all States constantly take steps in order to guarantee non-discrimination in access to economic, social and cultural rights everywhere and to ensure the minimum essential levels of each right enshrined in the covenant?[66] It should also be investigated to what extent national obligations of environmental law might be applicable internationally. Before legislating emission caps, States could be asked to do an impact assessment not only focusing on national (and border) regions, but also take into account consequences on a global level.[67]

Accepting the human rights-based approach to climate change, it is again within the competence of human rights institutions to deal with these questions. They should firstly concretize the State obligation to provide assistance to affected States and to co-operate internationally in reducing the effects of climate change on the enjoyment of human rights. Secondly, human rights institutions should monitor State compliance in this regard. During the State reporting procedure at the Human Rights Council as well as within the reporting systems of

64 See Limon, 'Human Rights Obligations and Accountability' (n 24) 578–581.

65 See Margit Ammer, 'Klimawandelinduzierte Umweltveränderungen als Ursache für freiwillige Binnenmigration: Ist Europa verpflichtet zu helfen?' (2010) 4 Zeitschrift für Menschenrechte 48, 62–64.

66 These criteria were established by the CESCR for the concretization of the national obligations to promote the ICESCR rights of a progressive nature; see CESCR, 'Operational Statement' (10 May 2007) UN Doc E/C.12/2007/1; see also Khaliq and Churchill (n 61) 213.

67 In favour of this approach Knox, 'Climate Change and Human Rights Law' (n 27) 201.

the human rights conventions, States should be invited to include information on how they fulfil their duty to provide international assistance to States most affected by the consequences of global warming.[68]

IV The Contribution of the Human Rights-Based Approach to the International Discourse on Climate Change

These conceptual approaches to climate change derived from international human rights law remain incomplete and it is even unclear whether the three questions could at all be answered in a way that satisfies human rights lawyers. At least the OHCHR's report on the relationship of climate change and human rights has not substantially changed the international discourse on how to manage the causes and consequences of global warming.[69] It is thus not very promising for the individual to claim compensation through climate change litigation procedures on the grounds of international human rights law. There is no consensus that human rights law explains the degree to which justice requires compensation for green house gas emissions.[70] Furthermore, States seemingly do not feel obliged to reduce emissions on the basis of their human rights obligations. Thus, it has to be concluded that at least in a formal sense contributing to climate change cannot be regarded a human rights violation.[71]

Is it therefore pointless to adopt a human rights-based approach to climate change – or does it even harm international human rights law by "cheapening the concept of rights"?[72] It may be stated that the human rights-based approach contributes to the international discourse on climate change in two significant ways: On a very practical level, conceptually linking climate change and human rights provides access to institutions. On a more theoretical level, the human rights-based approach to climate change introduces a vertical level to international climate law and refocuses it on the needs and subjective rights of individuals.

---

68 On reporting procedures see Walter Kälin, 'Examination of State Reports' in Helen Keller and Geir Ulfstein (eds), *UN Human Rights Treaty Bodies: Law and Legitimacy* (Cambridge University Press 2012) 37.

69 See Knox, 'Linking Human Rights and Climate Change' (n 5) 496; for some exemptions see McInerney-Lankford, Darrow and Rajamani (n 4) 55.

70 See Efthymiou (n 28) 120–124.

71 See Müller and Franzen (n 12) 24.

72 Kenneth Roth, 'Is climate change a human rigts issue?' bigthink.com/videos/re-is-climate-change-a-human-rights-issue-2 accessed 7 June 2013; a different approach: Edward Cameron, 'Human Rights and Climate Change: Moving from an Intrinsic to an Instrumental Approach' (2010) 38 Ga J Int'l Comp L 673.

1. Access to Institutions

Firstly, treating climate change as a human rights-related problem provides access to international institutions. Human rights bodies have proven to be forums of lively debates and expertise. This could be rewarding for the discussion of the human rights-related aspects of climate change as well.

The pertinent institutions of the international climate framework are mostly run by scientists and diplomats.[73] There is not much place for elaborating on the impacts of global warming.[74] The climate regime further does not establish a responsibility, *id est* liability framework, for individual claims against States for damages related to global warming and neither does it provide for institutions to which the individual may petition.[75] By linking climate change to human rights, a multitude of institutions become available as platforms for discussion. The Human Rights Council and the convention committees then are the forums where States as well as international experts negotiate problems of climate change, analyse them from a human rights angle and elaborate on related legal and political aspects.

With general comments or during the individual complaints procedure, the Human Rights Committee and the Committee on Economic, Social and Cultural Rights might investigate the effects of global warming on single convention rights. During the reporting procedures they could include climate change aspects and ask to what extent States co-operate internationally in order to mitigate the effects of climate change in Developing and Small Island States. NGOs could complement the process by presenting their points of view.

Within the United Nations system, it is furthermore the Human Rights Council's task to monitor emerging problems such as climate change "in order to ensure that preventive approaches are implemented to head off gross human rights violations".[76] Here also State representatives have the opportunity to illustrate

73 See e.g. for the Kyoto-Protocol Meinhard Doelle, 'Compliance and Enforcement in the Climate Change Regime' in Erkki J Hollo, Kati Kulovesi and Michael Mehling (eds), *Climate Change and the Law* (Springer 2012) 166, 169.

74 "One of the key failings of climate change diplomacy over the past two decades is that the phenomenon has been viewed as a scientific projection," Marc Limon, 'Human Rights and Climate Change: Constructing a Case for Political Action' (2009) 33 Harv Envt'l L Rev 439, 451.

75 Koivurova (n 41) 278–282; but see on some participation rights Heike Walk, *Partizipative Governance: Beteiligungsformen und Beteiligungsrechte im Mehrebenensystem der Klimapolitik* (Verlag für Sozialwissenschaften 2008) 153–170.

76 See Bertrand G Ramcharan, *The UN Human Rights Council* (Taylor and Francis 2011) 81.

how they are affected by climate change, share experiences on common climate change-related human rights problems and present their mitigation and adaptation strategies. During the Universal Periodic Review, for example, Vanuatu's representatives already made reference to the implications climate change had on the country's inhabitants.[77] He illustrated that especially women and children suffer from the consequences of extreme weather events and how the Government designs human rights-compliant response measures to adapt to climate change. Also, the State representative asked for international assistance in order to combat the consequences of climate change in his country.[78]

Thus, the charter-based as well as the treaty-based human rights institutions become a place where State efforts in combating climate change are put to a test from a human rights angle. One might argue that by 'burdening' human rights institutions with a climate change discussion the system is threatened by spreading itself too thin. This might actually be a realistic concern; however, it is rather an argument for allocating more resources to human rights institutions than against adopting a human rights-based approach to climate change. Linking human rights and climate change provides the international discourse on global warming with the necessary forums that finally moves the individual and his/her rights to the centre of the discussion.

2. Introduction of Subjective Rights

This leads to the second contribution of the human rights-based approach to the international discourse on climate change: It might add a vertical dimension of State obligations and introduces the possibility of subjective rights to international climate law. The pertinent international rules on climate change derived from international environmental law as well as the UNFCCC, including the Kyoto-Protocol, oblige States on a horizontal level.[79] Thus, States have obligations, e.g. the reduction of emissions, primarily towards other State parties. In contrast, human rights obligations function vertically; here an individual

77 HRC, 'Universal Periodic Review Highlights 12 May 2009 (morning, not an official record)' www.ohchr.org/EN/HRBodies/UPR/Pages/Highlights12May2009AM.aspx accessed 11 July 2013; see for a review of the innovative mechanisms of the Human Rights Council Rosa Freedman, *The United Nations Human Rights Council: A Critique and Early Assessment* (Taylor and Francis 2013) 254, 282.

78 HRC, 'Universal Periodic Review Highlights 12 May 2009' (n 77).

79 On the legal structure of the UNFCCC and the Kyoto-Protocoll as well as the contained obligations see Verheyen (n 43) 44, 108.

is entitled to a certain act or omission by a State.[80] Thus, the human rights-based approach could add to international climate law a new direction of obligations.[81] It at least re-centres the discussion on human needs and introduces the possibility of individual claims.

At present it is rather difficult to derive such a vertical structure from the UNFCCC.[82] It stands to reason, from a classical inter-state treaty with a horizontal structure, that subjective rights can also be derived and to what extent a human rights-oriented interpretation could assist with this endeavour. The general possibility of the existence of subjective rights in international law was recognized by the International Court of Justice in the *LaGrand*-case.[83] However, so far, there is not much consensus with regard to the identification of such rights.[84] It might be interesting to analyse Article 4 Paragraph 1 lit i in conjunction with Article 6 UNFCCC in this regard. By these provisions, State parties commit to promote education and training as well as public awareness on climate change and its effects and to encourage the widest possible participation in this process. Article 6 ensures public access to information on climate change and its consequences as well as public participation in developing adequate responses. These provisions could constitute independent subjective rights of individuals. A treaty interpretation in this regard would add a normative direction to international State obligations, especially for those parties to the UNFCCC that have not ratified the pertinent human rights conventions containing similar information and participation rights.[85]

---

80 See Kälin and Künzli (n 14) 90; Daniel Bodansky, 'Climate Change and Human Rights: Unpacking the Issues' (2010) 38 Ga J Int'l Comp L 511, 516; see however on some similarity of the two regimes Helen Keller and Geir Ulfstein, 'Introduction' in Helen Keller and Geir Ulfstein (eds), *UN Human Rights Treaty Bodies: Law and Legitimacy* (Cambridge University Press 2012) 2.

81 See Ammer (n 65) 59; likewise also Margaux J Hallt and David C Weiss, 'Avoiding Adaptation Apartheid: Climate Change Adaptation and Human Rights Law' (2012) 37 YJIL 309, 344.

82 See on human rights under the UNFCCC Limon, 'Human Rights Obligations and Accountability' (n 24) 582–586.

83 ICJ, *LaGrand Case* (Germany v USA) (2001) ICJ Rep 466, Paragraph 77.

84 See Anne Peters, 'Das subjektive internationale Recht' (2011) 59 Jahrbuch des öffentlichen Rechts der Gegenwart 411, 439.

85 See on this group of rights in the climate change context Svitlana Kravchenko, 'Procedural Rights as a Crucial Tool to Combat Climate Change' (2010) 38 Ga J Int'l Comp L 613.

Thus, human rights law could help "fill important gaps in the existing international climate change regime".[86] Accepting human rights law not as an isolated branch of international law, but as interacting with other fields of international law,[87] might significantly influence special regimes such as climate law by refocusing some of its rules on the needs and interests of the individual. In the OHCHR's view – although under the protest of industrialized States such as Canada[88] – human rights complement the UNFCCC by underlining that "the human person is the central subject of development and that international cooperation is not merely a matter of the obligations of a State towards other States, but also of the obligations towards individuals".[89] To what extent this approach actually introduces subjective rights to the international climate regime is a matter of the further development of international law. Generally speaking, it comes down to the question of what role the individual is to play in the international system.[90]

V Conclusion

It might be concluded that the United Nations as well as the human rights institutions have not yet fully explored the advantages of a human rights-based approach to climate change. There so far is no satisfying answer to the question of how climate change can, in a meaningful way, be conceptually connected to the normative legal construction of correlating rights and duties of international human rights law.

I suggest that international climate law benefits from a human rights-based approach not only by gaining a 'human face' but also from the fact that the human rights framework is more legalistic than environmental treaties. By approaching climate change from a human rights perspective, the discussion moves from the political arena to the area of international law.[91] For example, the reduction of green house gas emissions presents itself within human rights law as a question about what States shall do based on existing obligations; in

86 Limon, 'Human Rights and Climate Change' (n 74) 455.

87 See e.g. Linos-Alexander Sicilianos, 'The Human Face of International Law: Interactions between General International Law and Human Rights' (2012) 32 HRLJ 1.

88 See the submission of Canada to the OHCHR Study cited in Limon, 'Human Rights and Climate Change' (n 74) 460 n 116.

89 (15 January 2009) UN Doc A/HRC/10/61 Paragraph 86.

90 See e.g. Kate Parlett, *The Individual in the International Legal System: Continuity and Change in International Law* (Cambridge University Press 2010).

91 See Bodansky (n 80) 515.

the international climate regime it is merely regarded as a question of political negotiations and voluntary reduction targets.[92] Arguing that the political playing field of multiple climate change management strategies would illegitimately be narrowed can hardly convince if one accepts the idea of universal, equal and inalienable human rights. How States use their margin of appreciation requires monitoring, and thus the relationship of human rights and climate change needs to be (re-) discovered by the pertinent human rights institutions.

In the end, adopting a human rights-based approach to climate change can be regarded as a manifestation of the increasing focus of international law on the individual. More and more areas of international law are exploited for individual interests, are made accessible for the private person or are complemented by subjective rights.[93] Acknowledging this overarching process of a "humanization of international law"[94] the relationship of climate change and human rights requires further assessment. Ultimately, what then lies at the heart of the discussion of climate change as one of today's major challenges to the international community is not only the State's obligation but also the individual's right.

Bibliography

Ammer, M, 'Klimawandelinduzierte Umweltveränderungen als Ursache für freiwillige Binnenmigration: Ist Europa verpflichtet zu helfen?' (2010) 4 Zeitschrift für Menschenrechte 48

Bodansky, D, 'Climate Change and Human Rights: Unpacking the Issues' (2010) 38 Ga J Int'l Comp L 511

Burkett, M, 'Climate Reparations' (2009) 10 Melb J Int'l L 509

Cameron, E, 'Human Rights and Climate Change: Moving from an Intrinsic to an Instrumental Approach' (2010) 38 Ga J Int'l Comp L 673

Cullet, P, 'Liability and Redress for Human-Induced Global Warming: Towards an International Regime' (2007) 43 Stan J Int'l L 99

Doelle, M, 'Compliance and Enforcement in the Climate Change Regime' in E J Hollo, K Kulovesi and M Mehling (eds), *Climate Change and the Law* (Springer 2012)

---

92 Ibid 517.

93 See Peters (n 84).

94 Antonio Augusto Cancado Trindade, 'International Law for Humankind: Towards a New Jus Gentium' (vol II) (2005) 316 Recueil des Cours 15, 19; term also used by Theodor Meron, *The Humanization of International Law* (Nijhoff 2006).

Doussa, J v, A Corkey and R Chartres, 'Human Rights and Climate Change' (2007) 14 Austl Int'l L J 161

Efthymiou, D, 'Climate Change, Human Rights and Distributive Justice' (2009) 20 Finnish YBIL 111

Faure, M G and A Nollkaemper, 'International Liability as an Instrument to Prevent and Compensate for Climate Change' (2007) 26 Stan Envt'l L J 123

Food First Information and Action Network Deutschland, 'Menschenrechte im Klimawandel – Anforderungen an die deutsche und internationale Klimapolitik 2009' www.fian-deutschland.de/online/index.php?option=com_remository&Itemid=160&func=startdown &id=165 accessed 7 June 2013

Ford, J D, 'Supporting Adaptation: A Priority for Action on Climate Change for Canadian Inuit' (2008) 8 Sustainable Dev & Pol'y 25

Freedman, R, *The United Nations Human Rights Council: A Critique and Early Assessment* (Taylor and Francis 2013)

Gordon, J, 'Inter-American Commission on Human Rights to Hold Hearing after Rejecting Inuit Climate Change Petition' (2007) 7 Sustainable Dev L & Pol'y 55

Grote, R, 'The Optional Protocol to the International Covenant on Social, Economic and Cultural Rights: Towards a More Effective Implementation of Social Rights?' in H P Hestermeyer et al (eds), *Coexistence, Cooperation and Solidarity, Liber Amicorum Rüdiger Wolfrum* (vol I, Nijhoff 2012)

Hallt, M J and D C Weiss, 'Avoiding Adaptation Apartheid: Climate Change Adaptation and Human Rights Law' (2012) 37 YJIL 309

Hollo, E J, K Kulovesi and M Mehling, 'Climate Change and the Law' in E J Hollo, K Kulovesi and M Mehling (eds), *Climate Change and the Law* (Springer 2012)

Innuit Circumpolar Council, 'Petition to the Inter-American Commission on Human Rights Seeking Relief from the Violations Resulting from Global Climate Warming Caused by Acts and Omissions of the United States' (7 December 2005) www.inuitcircumpolar.com/files/uploads/icc-files/FINAL PetitionSummary.pdf accessed 6 June 2013

Kälin, W, 'Examination of State Reports' in H Keller and G Ulfstein (eds), *UN Human Rights Treaty Bodies: Law and Legitimacy* (Cambridge University Press 2012)

– and J Künzli, *Universeller Menschenrechtsschutz* (2nd ed, Helbing-Lichtenhahn 2008)

Keller, H and G Ulfstein, 'Introduction' in H Keller and G Ulfstein (eds), *UN Human Rights Treaty Bodies: Law and Legitimacy* (Cambridge University Press 2012)

Keller, H and L Grover, 'General Comments of the Human Rights Committee and Their Legitimacy' in H Keller and G Ulfstein (eds), *UN Human Rights Treaty Bodies: Law and Legitimacy* (Cambridge University Press 2012)

Khaliq, U and R Churchill, 'The Protection of Economic and Social Rights: A Particular Challenge?' in H Keller and G Ulfstein (eds), *UN Human Rights Treaty Bodies: Law and Legitimacy* (Cambridge University Press 2012)

King, K, 'Climate Change and the Inuit: A Melting of Actions into a Cloudy Mess' (2009) 17 Se Envt'l L J 481

Klein, E, 'Self-Contained Regime' in R Wolfrum (ed), *The Max Planck Encyclopedia of Public International Law* (2008 online ed, accessed 6 July 2013)

Knox, J, 'Climate Change and Human Rights Law' (2009) 50 Va J Int'l L 163

– 'Linking Human Rights and Climate Change at the United Nations' (2009) 33 Harv Envt'l L Rev 477

Koivurova, T, 'International Legal Avenues to Address the Plight of Victims of Climate Change: Problems and Prospects' (2007) 22 J Envt'l L & Litig 267

Kravchenko, S, 'Procedural Rights as a Crucial Tool to Combat Climate Change' (2010) 38 Ga J Int'l Comp L 613

Limon, M, 'Human Rights and Climate Change: Constructing a Case for Political Action' (2009) 33 Harv Envt'l L Rev 439

– 'Human Rights Obligations and Accountability in the Face of Climate Change' (2010) 38 Ga J Int'l Comp L 543

McInerney-Lankford, S, M Darrow and L Rajamani, *Human Rights and Climate Change: A Review of the International Legal Dimensions* (World Bank Publications 2011)

Meron, T, *The Humanization of International Law* (Nijhoff 2006).

Müller, C and K Franzen, 'Der Klimawandel und das Menschenrechtssystem der Vereinten Nationen' (2010) 4 Zeitschrift für Menschenrechte 7

Parlett, K, *The Individual in the International Legal System: Continuity and Change in International Law* (Cambridge University Press 2010)

Peters, A, 'Das subjektive internationale Recht' (2011) 59 Jahrbuch des öffentlichen Rechts der Gegenwart 411

Posner, E A, 'Climate Change and International Human Rights Litigation: A Critical Appraisal' (2007) 155 U Pa L Rev 1925

Ramcharan, B G, *The UN Human Rights Council* (Taylor and Francis 2011)

Rodely, N S, 'UN Treaty Bodies and the Human Rights Council' in H Keller and G Ulfstein (eds), *UN Human Rights Treaty Bodies: Law and Legitimacy* (Cambridge University Press 2012)

Sicilianos, L A, 'The Human Face of International Law: Interactions between General International Law and Human Rights' (2012) 32 HRLJ 1

Stallard, H, 'Turning up the Heat on Tuvalu: An Assessment of Potential Compensation for Climate Change Damage in Accordance with States Responsibility under International Law' (2009) 15 Canterbury L Rev 163

Trindade, A A C, 'International Law for Humankind: Towards a New Jus Gentium' (vol II) (2005) 316 Recueil des Cours 15

Ulfstein, G, 'Individual Complaints' in H Keller and G Ulfstein (eds), *UN Human Rights Treaty Bodies: Law and Legitimacy* (Cambridge University Press 2012)

Verheyen, R, *Climate Change Damage and International Law: Prevention Duties and State Responsibility* (Nijhoff 2005)

Voigt, C, 'State Responsibility for Climate Change Damages' (2008) 77 NJIL 1

Walk, H, *Partizipative Governance: Beteiligungsformen und Beteiligungsrechte im Mehrebenensystem der Klimapolitik* (Verlag für Sozialwissenschaften 2008)

Impacts of Climate Change on Cultural Heritage and Cultural Diversity

Michael Turner & Rachel Singer*

Urban Resilience in Climate Change

Abstract The term Urban Resilience re-emerged at the beginning of the 21st century as a counterterrorism policy. In recent years it has taken on new policy meanings and has broadened to formulate planning solutions that anticipate a wide range of potential disasters, natural and human, many of which are associated with climate change and extreme weather conditions. There are different methods and approaches on how to best meet the challenges, though mitigation and adaptation are both needed to cope with the predicted scenarios. A cross-disciplinary approach and a re-examination of existing heritage policies in relation to Urban Resilience and climate change will allow more flexibility to cope with these contemporary and future threats. This paper, within the framework of the UNESCO Recommendation on Historic Urban Landscapes, proposes new integrative conceptual tools to understand Urban Resilience and stresses the importance of heritage and cultural diversity in the discussion, thus ensuring sustainability between communities and preserving heritage for future generations.

I What is Urban Resilience?

Resolving the aftermaths of sudden changes has been an agenda item from time immemorial. The ancients developed narratives where the elements of fire and water were evoked to explain disasters, and the floods of Gilgamesh and Noah and the fire and brimstone of Lot were just the tip of the iceberg. Aristotle in

* Prof. Michael Turner holds the UNESCO Chair in Urban Design and Conservation Studies at Bezalel Academy of Art and Design in Jerusalem, Israel, coordinated the UNESCO World Heritage Second International Workshop on Disaster Risk Reduction to Cultural Heritage in Acre and peer reviewed the publication on Managing Disaster Risks for World Heritage and participated in the preparation of Disaster Risk Management of Cultural Heritage in Urban Areas. He has accompanied the UNESCO Recommendation on Historic Urban Landscapes since its inception in 2005.
Rachel Singer lives in Jerusalem, Israel, and is currently working towards a Graduate degree in Urban Design at the Bezalel Academy of Art and Design. Previous studies include Conservation of Material Heritage (M.A., Haifa University) and Geography (B.A., Hebrew University).

ancient Greece categorizes disasters as the result of natural phenomena and not of supernatural interventions. However, more important is not the explanation of their occurrence but the social implications of recovery, and the capacities for moving ahead; in short, resilience. The concept of 'resilience' needs to be seen as a panacea to the negative impacts of change, whether gradual or sudden, whether man-made or though 'natural disasters', and as a proponent of the positive effects of growth.

Resilience combines the different periods of disaster cycles, including: mitigation, preparedness, response, recovery and adaptation.[1] In the last decade the rhetoric of resilience and its underpinnings have slowly infused a variety of policymaking circles and a wide range of disciplinary areas,[2] being more inclusive by relating to both the physical and social components of the city.[3] Urban resilience is a proactive approach that provides both a lens through which we can analyse the different problems that cities face as well as providing a framework with space for solutions. Several phases in the evolution of urban resilience policies have been noted, initially reactive, expanding to take preventive actions and then followed by an attempt to incorporate aspects of resilience into daily routine, through the anticipation of shocks on multiple levels. The current approach sees the path to resilience as inclusive, incorporating the individual, institutions and neighbourhoods in order to achieve the broadest possible form of participation.[4]

Based on the three pillars of sustainability and extending from the physical infrastructures to include the socio-economic and environmental, the United Nations Office for Disaster Risk Reduction (UNISDR) broadened the definition of resilience as:

> "[...] the ability of a system, community or society exposed to hazards to resist, absorb, accommodate to and recover from the effects of a hazard in a timely and efficient manner, including through the preservation and restoration of its essential basic structures and functions".[5]

---

1 Geoff O'Brien and Paul Read, 'Future UK Emergency Management: New Wine, Old Skin?' (2005) 14 Disaster Prevention and Management 353.

2 Jon Coaffee, 'Rescaling and the Responsibilising of the Politics of Urban Resilience: From National Security to Local Place Making' (2013) 33 Politics 240.

3 Jon Coaffee and Lee Bosher, 'Integrating Counter-terrorist Resilience into Sustainability' (2008) 161 Proceedings of the ICE – Urban Design and Planning 75.

4 See Coaffee (n 2) 243, 246–248.

5 UNISDR, 'Terminology: Resilience' (The United Nation Office for Disaster Risk Reduction, 30 August 2007) www.unisdr.org/we/inform/terminology accessed 21 January 2014.

As it is often not possible to prevent disasters, governments and NGOs are aiming to increase the resilience of communities and societies through a combination of research, policy and programme development as well as through management and education. Social resilience to disasters is associated with a number of characteristics, including: *resistance*, *recovery* and *creativity*.

Resistance refers to the magnitude of disruption that can be absorbed by a community before undergoing significant changes. *Recovery* refers to the community's ability to mend following a disaster and its return to function. The recovery can be measured as the time needed to rebound to levels similar to those before the disaster. Binding this together is social resilience as a function of multiple external factors as well, including environmental, political and economic. These forces also have a significant impact on the recovery, for instance strong economic forces can buoy a relatively quick recovery while a sluggish economic climate can hinder. Reaching the previous state of equilibrium post disaster is not the final stage but a point of reference in the process of continuing change. It is possible to learn from disasters and utilize these situations as opportunities to springboard further preparedness and resilience for future events; this is the characteristic of *creative* communities. Even within communities that have a high level of resilience there are groups with fewer tools to cope with disasters and tend to be disproportionately impacted; including women, children, the elderly and low-income groups.[6] These particular characteristics of social behaviour need to be studied to achieve a more durable approach to disasters by inculcating day-to-day reactions to 'everyday' incidents.

What happens to cities after disasters? Disasters, strangely enough, need to be understood also as challenges and opportunities.[7] This is where the city and its physical and social structures excel. Being an inherent part of urban resilience, it is a system that is able to absorb shocks, and the higher the level of resilience the greater the shock that that can be absorbed, and vice versa. In the outline of their book on the resilient city, Campanella and Vale describe the city as a phoenix, able to regenerate from the ashes of destruction; it is the exception to the rule that cities are lost.[8]

6 Brigit Maguire and Patrick Hagan, 'Disasters and Communities: Understanding Social Resilience' (2007) 22 Australian Journal of Emergency Management 16.

7 Michael Turner, 'Disaster preparedness and the aftermath: Risk Management as a component of conservation theories for Historic Urban Environments/Landscapes', Presentation at the workshop on sustainable protection of Cultural Heritage in earthquake zones (Kyoto 30 August 2009, on file with author).

8 Lawrence J Vale and Thomas J Campanella, *The Resilient City: How Modern Cities Recover from Disaster* (Oxford University Press 2005).

Contemporary research on climate change during the relatively short period of Anatomically Modern Human existence explores the link between climate disasters and the decline of urban culture in major cities, highlighted during the mediaeval period in Baghdad and Constantinople, which suffered from periods of extreme cold and drought. The effects of disasters on the populations in these cities and the resilience of the ruling administrations was the result of the intensity of the calamity and the level of preparedness (or unpreparedness), specifically how much food had been stored during periods of abundant yields.[9] There are examples of coping mechanisms illustrating different levels of mitigation and adaptation dating back thousands of years. One might also conjure up the strategic policy of the biblical Joseph, chief advisor to Pharoah, as early documented evidence for risk preparedness. In anticipation of a seven year drought, Joseph proposed a massive plan to store food to feed the Egyptian populace.

II Measuring Urban Resilience

As the term resilience is still surrounded by a certain ambiguity regarding its precise meaning and application, there is a present hazard that it may morph into a type of fashionable slogan lacking applicable content[10] or alternatively develop into manageable urban management strategies. The measurement of resilience could benefit from von Bertalanffy's General System Theory and its applications to urban processes of change in the city. The decision-making systems of natural and biological organisms resilience identifying feed-back[11] and feedforward processes (*Figure 1*) can be referred to in the analogies of the resilience of city as 'design by error' and 'zero error in design'.

In feedback processes adaptive systems are programmed to self regulate their responses to the surrounding environment. Information regarding various situations is channelled to points in the system that determines the appropriate response.[12] Resilience, in essence, is the development of feed-back processes in the

9 Ronnie Ellenblum, *The Collapse of the Eastern Mediterranean: Climate Change and the Decline of the East, 950–1072* (Camebridge University Press 2012) 5, 22, 88.

10 Adam Rose, 'Economic Resilience to Natural and Man-made Disasters: Multidisciplinary Origins and Contextual Dimensions' (2007) 7 Environmental Hazards: Human and Social Dimensions 1.

11 Ludwig von Bertalanffy, *General System Theory: Foundations, Development, Applications* (Braziller 1968) 42–44.

12 Dodd H Bogart, 'Feedback, Feedforward, and Feedwithin: Strategic Information in Systems' (1980) 25 Behavioral Science 238.

face of uncertainty by defining points on the process as barriers of irreversibility (*Figure 2*). These barriers of limits will determine the scale of the activities and the costs needed for 'undoing' the actions. These uncertainties might be understood within decision-making processes using feed-back urban processes and dynamic management as opposed to feed-forward actions based on a more static master-plan project approach. Scenario planning is currently replacing the *psychedelic* Master Plan with zoning thus providing a link to the risk management of the city, in addition to providing for a culture of resilience and a live training facility for stakeholders.

Figure 1: Diagram illustrating concepts of "Feedback" and "Feedforward", based on von Bertalanffy, the upper diagram shows Control by Error - the feedback mechanism allowing the existence of an error if it is reversible, and the system to incorporate adjustments. The lower diagram, Control by Design-shows the feedforward mechanism, a system that constantly strives to maintain a state of equilibrium; it allows zero error and is not adaptable to changes (Turner)

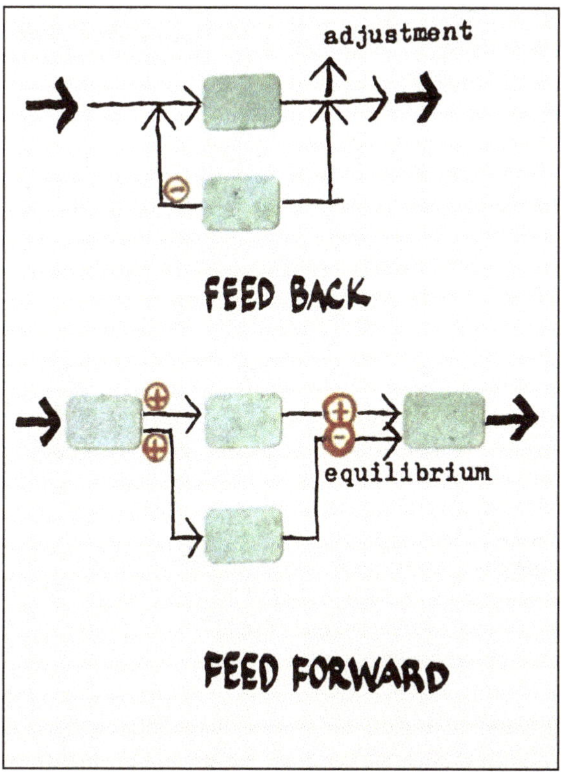

Figure 2: Diagram illustrating the concept of "Barriers of Irreversibility", within a single stage there is room for error and feedback, however once the barrier has been passed entering the feed-forward range, it is not possible to reintroduce the feedback mechanism (Turner)

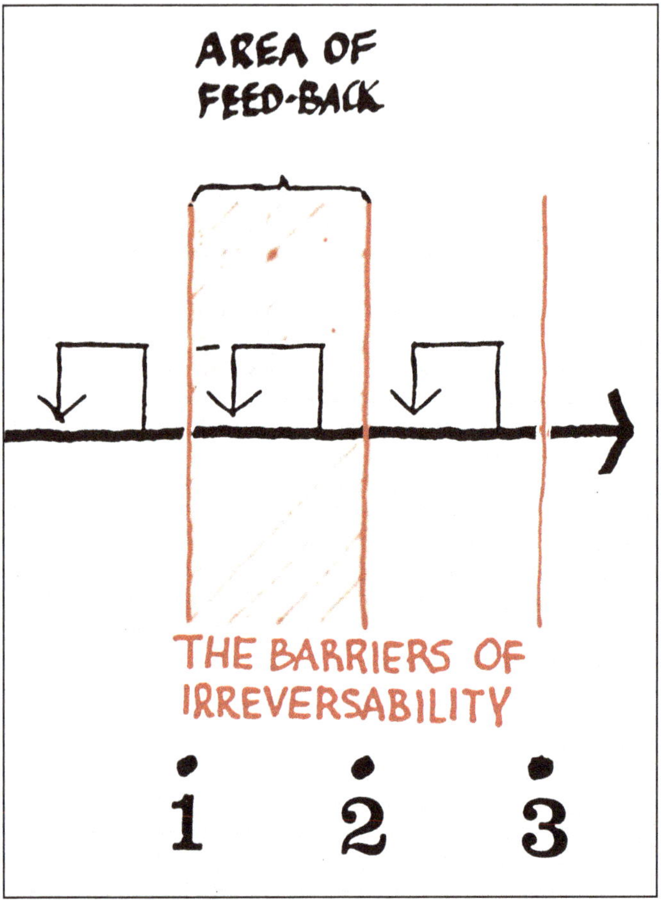

Not all errors can be tolerated, and due to the dynamic structure of the city, the boundaries of irreversibility must be recognised as 'points of no return'. Actions and reactions must have limits and these might be defined as a reiterative process contained by a defined set of parameters, in which feedback can be applied to the process.

Within this planning process it is possible to identify a range of resilience that is bordered by limits on the one hand and design on the other (*Figure 3*).

Figure 3: Diagram on the left explains the links between the planning process, the design process and imposed limits using a pendulum analogy, the design and adaptation oscillates between the plan and its limits, through different "zones" to find the correct balance – thus identifying the range of resilience. The diagram on the right differentiates between natural design processes, which feature an S curve remaining within limits and planned processes which will exceed them leading to growth beyond limits (Turner)

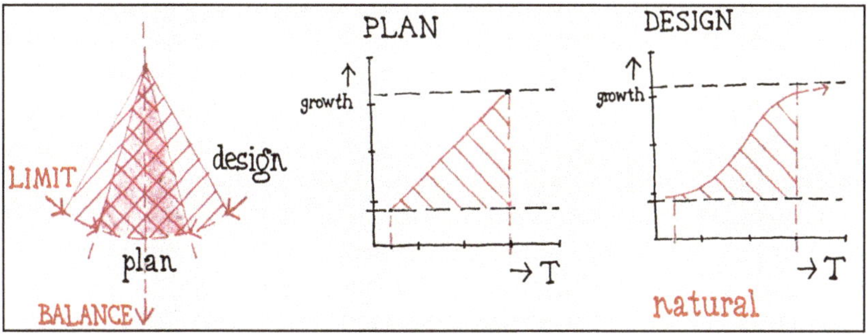

III Increasing Urban Resilience as a Response to Climate Change

Reports from the Intergovernmental Panel on Climate Change show that climate change is a factor in the increasing trend of extreme weather events, ranging from floods and droughts to cyclones and extreme temperatures.[13] While climate change is often perceived as a long term effect, not to be reckoned with during the politician's fickle life span, a more immediate phenomenon seems to be the appearance of severe climatic activities occurring in a series of 'short sharp shocks'. An example is an unusual desert flash storm that washed away conservation works at Masada, Israel, precipitating a need to nationally redefine the mortar components of the bonding limes for conservation. Similar flash storms have created landslips and mudslides in areas of development with marginalised communities in Italy, Central and Latin America. At an urban scale, two drought seasons in the rural hinterland of Harar, Ethiopia has brought about a massive migration to the city with squatter neighbourhoods appearing on the urban fringe. This in turn has created new pressures and challenges for the resilience of the city.

13 IPCC, Managing the Risks of Extreme Events and Disasters to Advance Climate Change Adaptation. A Special Report of Working Groups I and II of the Intergovernmental Panel on Climate Change (Oxford University Press 2012) 29.

Climate change is an integral part of the Earth's history. The debate on the speed and characteristics of these changes that might be attributed to human activities does not affect the necessity of identifying the two primary methods in use to combat the ensuing urban effects; that of mitigation and adaptation.

Mitigation is referred to as the lessening or limitation of the adverse impacts of hazards and related disasters. The adverse impacts of hazards often cannot be fully prevented, though their scale or severity can be substantially lessened by various strategies and actions. Mitigation measures encompass engineering techniques and hazard-resistant construction as well as improved environmental policies and public awareness.[14] These measures primarily focus on the long term and can have both international and local applications.[15]

Adaptation is "the adjustment in natural or human systems in response to actual or expected climatic stimuli or their effects, which moderates harm or exploits beneficial opportunities".[16] Adaptation is generally on a smaller scale, referring to the impacted system and often site specific.[17] The applicable dimensions of resilient urban infrastructure are currently being examined and extended to the socioeconomic and environmental conditions, resulting in methodologies as Social Ecological System (SES).[18]

At present, efforts are aimed to create procedures that are applicable during both normal and disaster incidents, some cities are promoting campaigns to the general public which link disaster risk reduction with sustainable living. The goal of these campaigns is to change individual, household and community behaviours and raise the level of resilience.[19]

At the global level, UN Habitat has embarked on a mission to create City Resilience Profiles, one of the current key problems identified being that there is a lack of an integrative urban systems approach, and many of the models belong to specific sectors. The need for interdisciplinary action was articulated by Richard

14 UNISDR, 'Terminology: Mitigation' (n 5).

15 Adriana Galderisi and Floriana F Ferrera, 'Enhancing Urban Resilience in Face of Climate Change: A Methodological Approach' (2012) 5 Journal of Land Use, Mobility and Environment 69.

16 UNISDR, 'Terminology: Adaptation' (n 5).

17 Galderisi and Ferrera (n 15).

18 James P Evans, 'Resilience, Ecology and Adaption in the Experimental City' (2011) 36 Transactions of the Institute of British Geographers 223.

19 UNISDR, 'Making Cities Resilient' (2012) www.unisdr.org/campaign/resilientcities/ accessed 23 January 2014.

Haigh, the director of the Salford University Centre's Disaster Mitigation and Reconstruction Programme:

> "The complex nature of disasters has led to recognition that risk reduction through increased resilience will require a strategy that is inter-disciplinary. True interdisciplinarity only occurs where a number of separate disciplines surrender their own concepts and goals, and collectively define themselves by reference to a common set of strategic concepts and goals".[20]

In parallel, the UNISDR report on "Making Cities Resilient" discusses the importance of international cooperation and knowledge sharing. The "City to City Sharing Initiative" between Makati City, Quito and Kathmandu included an exchange of good practices in managing and reducing urban risk through improved land use planning, emergency management and increased public awareness. Preservation and rehabilitation strategies to ensure the protection of Cultural Heritage were also shared.[21] Best cases of resilience occur when the formula for managing an emergency or disaster is paralleled in normal situations and day-to-day operations. It is clear that international cooperation provides a wider perspective of the city and its context. Further, with relevant financial support cooperation can provide the resources for the mitigation of disasters on a global scale, but it is the best practice in the tools of adaptation where person-to-person cooperation can take place and that actions are really needed. This has been highlighted in the UN Habitat report and the case studies indicating the sharing of knowledge on managing systems in a sustainable way. The seventh framework action on Designing Safer Urban Spaces (DESURBS) goes a long way in establishing a database and proposing methodologies and protocols that promote urban resilience.[22]

In planning urban forms it is vital to use resources to achieve multiple goals which factor in, as much as possible, both mitigation and adaptation. Buildings should integrate renewable power sources, such as solar energy, and factor in surrounding climatic conditions to enable a comfortable internal environment.[23] At the same time these buildings might form part of an architectural conservation area using existing sustainable historic buildings or benefit from policies of reuse and recycling of the urban fabric.

---

20 UN Habitat, 'City Resilience Profiling Programme: Providing national and local governments with tools for measuring and increasing resilience to multi-hazard impacts including those associated with climate change' (UN Habitat information brochure 2012) www.unhabitat.org accessed 23 January 2014, 2.

21 UNISDR, 'Making Cities Resilient' (n 19).

22 See DESURBS, 'Project' desurbs.eu/index.php/project accessed 23 January 2014.

23 Elisabeth M Hamin and Nicole Gurran, 'Urban Form and Climate Change: Balancing Adaptation and Mitigation in the U.S. and Australia' (2009) 33 Habitat International 242.

To demonstrate the interlacing of urban components and attributes of the varied actions for mitigation and adaptation, Ewing, Bartholomew, Winkelman, Walters & Chen formulated a mitigation policy goal of reducing car emissions through the measurement of Vehicle Miles Travelled (VMT). They identify five Urban *Ds*:

- *Density* – higher persons, jobs and/or dwelling units per unit area
- *Diversity* – greater mix of land uses to include residential, employment, and retail/services in close proximity to each other
- *Design* – smaller block size or larger number of intersections per square mile, more sidewalk coverage, smaller building setbacks, smaller street width, more pedestrian crossings, more street trees
- *Destination* – more jobs or other attractions are reachable within a reasonable travel time, tends to be highest in urban cores
- *Distance to Transit* – shorter distance from home or work to the nearest rail station or bus stop[24]

While this might be perceived as very specific and particular, the parameters or attributes clearly have wider implications and can be applied usefully as attributes for other urban policies, such as crime prevention, risk preparedness, environmental hazards and socio-economic equity within a comprehensive resilience policy.

The following matrix examines the 5 *Ds* and their urban attributes in relation to potential disasters that have been linked with extreme weather incidents associated with climate change – flooding, landslides, storms and extreme temperatures. This matrix could be developed for other urban attributes and is important for the consideration of the integrative process.

Table 1: The 5 Ds and their urban attributes in relation to potential disasters

	mitigation	**preparedness**	**response**	**recovery**	**adaptation**
Density	3	1	1	1	2
Diversity	1	3	2	3	3
Design	3	3	1	3	3
Destination	1	1	1	1	2
Distance to Transit	1	1	3	2	1

A scale with a rating between 1 (low correlation) and 3 (high correlation) was used to demonstrate the relationship between climate related disasters and urban components. The rating could be allocated by different stakeholders to achieve coordinated actions

24 See Reid Ewing et al, *Growing Cooler: Evidence on Urban Development and Climate Change* (Urban Land Institute 2008) 70–71.

The matrix might be expanded to include other components of resilience, such as translating *Distance to transit* to the *Distance to first aid and assistance.* Another factor that has yet to be harnessed within the urban scene is the potential of the *smart city* and the applications of ICT in the management of the city whereby bottle-necks and threats can be managed efficiently. This potential needs to be tapped and can possibly support both mitigation and adaptation.

Local actions for urban adaptation are the key to resilience and its integrative approach can respond to climate change, both incremental and extreme conditions. The main sectors that were identified in relation to climate change include: energy, transportation, water management, natural hazards, waste management, planning and governance.[25] Urban resilience will be most effective when these sectors are integrated with the 5 Ds for mitigation.

All this highlights the need for strengthening decision-making processes to include resilience, which could be effective through the expansion of the Environmental Impact Assessment process with specific references to Risk Management rather than an independent stand-alone document. This should underscore the processes that develop feed-back and feed-forward recommendations and conclusions based on these assessments and their monitoring utilising tools as Results Based Management to evaluate the true added value.

IV Heritage as a Contributor to Urban Resilience

The open secret for urban resilience is in the mixed uses of the city, its cultural diversity and sensitivities to environmental assessment. While the latest UNESCO recommendation on the Historic Urban Landscape underscores the need for an integrative approach to planning, these 'mainstream' actions need to be expanded, as most urban development affected by climate change and other disasters are usually in areas that are environmentally unsustainable on 'left-over urban spaces'[26] that were cheap and readily available. Too often it is the underprivileged that are affected by incidents in the city that involve urban disasters. Hurricane Katrina in 2005 is a prime example that links the needs for social equity with climate change, where the lowest socio-economic groups have been affected.

25 World Bank, 'Cities and Climate Change: An Urgent Agenda' (2010) 10 Urban Development Series Knowledge Papers 6.

26 The term 'space left over after planning' (SLOAP) was coined by Prof Leslie Ginsburg of the Architectural Association in London during the 1960's.

The Low Elevation Coastal Zone (LECZ), defined by McGranahan, Balk & Anderson is "the contiguous area along the coast that is less than 10 metres above sea level that covers 2 per cent of the world's land area but contains 10 per cent of the world's population and 13 per cent of the world's urban population".[27] It is not surprising that small increases in sea levels can have devastating effects. The aqua alta of Venice, the Thames estuary, the sea towns of Zanzibar and Lamu in East Africa, the river settlements in Bangkok, Jakarta and Accra all share similar challenges, and notably the flooding of low lying areas as in Bangladesh. Sea walls, barriers and breakwaters have become immediate necessities, these solutions, in many cases, are not sustainable in the long term and are not economically or environmentally viable. In addition, they often affect the city's historic structures within the wider urban fabric. More balanced planning utilising open spaces can increase urban liveability on one hand and help cope with flooding on the other. Good environmental planning solutions are preferable to expensive man-made engineering options.

Both Cultural and Natural Heritage are increasingly affected by events that result from human and natural causes. Heritage sites provide a significant element of social cohesion and sustainable development and their erosion has extremely negative ramifications for society and its resilience. Human behaviour and decisions developed in normal times can greatly increase the resilience of these sites, thus reducing the impact of potential disasters. Specific disasters are generally far more destructive than prolonged processes of deterioration. A short-lived disaster can swiftly and suddenly devastate an entire area, and therefore managing disaster risks is an urgent issue:

> "Heritage if well maintained can positively contribute to reducing disaster risks. This is true not only for natural heritage resources that guarantee the proper functioning of ecosystems and the beneficial effect of their goods and services, but also for cultural heritage properties that- as a result of traditional knowledge accumulated over centuries have proven to be resilient to disasters while providing shelter and psychological support to affected communities".[28]

Where traditional knowledge systems exist it is vital to integrate them into the Disaster Risk Management.[29] These wisdoms and memories of the city can

27 Gordon McGranahan, Deborah Balk and Bridget Anderson, 'The Rising Tide: Assessing the Risks of Climate Change and Human Settlements in Low Elevation Coastal Zones' (2007) 19 Environment and Urbanization 17.

28 UNESCO, 'Managing Disaster Risks for World Heritage' (2010) World Heritage Resource Manual whc.unesco.org/document/104522 accessed 23 January 2014, 3–4.

29 Ibid 40.

provide information and awareness, enabling heritage to contribute to the development of resilience parameters.[30] A World Heritage workshop held in Acre, Israel in 2009 continued to expand on the 2008 results established at the Olympia meeting in the field of Disaster Risk Reduction (DRR), emphasising the policy document on the Impact of Climate Change on World Heritage Properties adopted by the General Assembly of the States Parties to the World Heritage Convention in 2007. The Acre Workshop further elucidated the link between events related to climate change and DRR.[31]

The annual UNESCO State of Conservation reports offer an indicator of the current threats to World Heritage properties. Of 1.007 (as of 2014[32]) listed properties, 250 are registered World Heritage Cities within the framework of the Organization of World Heritage Cities (OWHC), and within the category of climate change and severe weather events, flooding is the greatest current challenge. The statistics of the World Heritage Centre show that climate and severe weather events are affecting 10 per cent of the properties and are interlinked to other factors mainly attributed to management (71 per cent) or building and development (43 per cent).[33]

In the fields of conservation there has been much recognition of the importance of cultural diversity – of place, people, time and narrative, notably with the Convention on the Protection and Promotion of the Diversity of Cultural Expressions.[34] Since the 1990's important texts have strived to integrate cultural diversity into the conservation process, recognising urban conflicts on one hand and the needs for their resolution through co-existence or compromise, on the other. The Nara Document (1994)[35] and subsequent Burra Charter (1999)[36]

---

30 Rohit Jigyasu et al, 'Heritage and Resilience' 4th Session of the Global Platform for Disaster Risk Reduction (2013) whc.unesco.org/document/122923 accessed 23 January 2014, 30.

31 UNESCO, 'Recommendations of the Second International Workshop on Disaster Risk Reduction to Cultural Heritage Acre (Israel)' (14 November 2009) whc.unesco.org/en/events/610 accessed 23 January 2014.

32 For dynamic reference see whc.unesco.org/en/list.

33 For dynamic reference see whc.unesco.org/en/activities/634/.

34 Adopted 20 October 2005, entered into force 18 March 2007, 2440 UNTS 311.

35 ICOMOS, 'The Nara Document on Authenticity' drafted at Nara Conference on Authenticity in Relation to the World Heritage Convention held at Nara, Japan, from 1–6 November 1994 www.icomos.org/charters/nara-e.pdf accessed 23 January 2014.

36 ICOMOS Australia, 'The Burra Charter: The Australia ICOMOS Charter for Places of Cultural Significance 1999' australia.icomos.org/publications/charters/accessed 23 January 2014.

together with the 2011 UNESCO Recommendation on Historic Urban Landscapes (HUL)[37] all strengthen the importance of integrative processes in the evaluation of urban challenges and needs. The new HUL Recommendation states that the

> "[...] approach goes beyond the preparations and readiness for disasters, but includes the day-to-day management of the city and the lives of the individuals, which make up its identity. Multi-layered spaces and multi-layered times are a key component in establishing urban resilience".[38]

Cultural diversity is an essential element in our society, the acceptance of 'the other', the enriching experiences of the society and more important their cultural expressions and manifestations in the urban fabric.

Urban resilience is achieved in many instances through the need to resolve conflicting dilemmas involving many socio-cultural groups. Therefore the resolution of conflict, by compromise or coexistence, as outlined in the Code of the Burra Charter can be made relevant, assuming that:

(i) the healthy management of cultural difference is the responsibility of society as a whole;
(ii) in a pluralist society, value differences exist and contain the potential for conflict; and
(iii) ethical practice is necessary for the just and effective management of places of diverse cultural significance.

It is important to highlight that more often than not, it is the intra and not intercultural conflicts that are critical and need resolution. These might be achieved through the management of the city that combines these three assumptions into a single policy. This concept is further highlighted in Article 13, which sets out the basic tenet that

> "[...] co-existence of cultural values should be recognised, respected and encouraged, especially in cases where they conflict. For some places, conflicting cultural values may affect policy development and management decisions".[39]

---

37 UNESCO, 'Recommendation on the Historic Urban Landscape' (10 November 2011) whc.unesco.org/en/activities/638 accessed 23 January 2014.

38 Ibid.

39 In this article the term 'cultural values' refers to those beliefs which are important to a cultural group, including but not limited to political, religious, spiritual and moral beliefs. This is broader than values associated with cultural significance.

Diversity must also be harnessed as an important tool to build resilience, as monoculture has inherent weaknesses, showing much less resilience than a diverse system that has multiple resources to cope with and recover from crises.[40] Diversity in general and cultural diversity in particular is an important concept to develop in support of resilience in all its component parts of sustainability, physical, social, environmental and economic and are essential in combating the urban effects of climate change.

V Conclusion

The significance of climate change lies in its interactions with other aspects of change and stress and its impacts should be considered within such a multi-cause context.[41] At first sight it might seem strange to link climate change with cultural diversity, equity and sustainability, but the parameters and attributes of each when overlaid show a close inter-relationship, strengthening the need for a multi-disciplinary approach. This links social inequality to the effects of urban disasters from climate change, where poorer communities find themselves in areas of weak environmental conditions sensitive to disasters thus demanding a more comprehensive analysis for applying the test of resilience to the city.

Cultural mapping as applied in the UNESCO Recommendation on HUL could generate a more integrated awareness of knowledge and traditions not only by the local community but by the public at large, thus understanding the spirit of the populace and place. The mapping of environmental weaknesses will be critical in developing an urban resilience policy, absorbing historic wisdoms and traditional knowledge, wherever the experiences of, for instance, low-lying slope areas, orientation and geo-morphology are known.

Responding to change is the challenge of the cities where social migrations, economic revolution become day-to-day occurrences. The smart city of the digital age offers the potential to connect systems and manage disaster risks based on a growing body of knowledge, now manageable at a macro-scale. The innovative changes in life-styles due to this digital revolution can re-organise existing land use patterns and provide mixed uses for urban resilience. The way forward is to develop an integrative methodology whereby language should transcend professions and disciplines in the spirit of the General System Theory; for instance

40 Carl Folke et al, 'Resilience and Sustainable Development: Building Adaptive Capacity in a World of Transformations' (2002) 31 Journal of the Human Environment 438.

41 IPCC, 'Climate Change and Water: Technical Paper VI' (June 2008) www.ipcc.ch/pdf/technical-papers/climate-change-water-en.pdf accessed 12 June 2013.

ruptures refer equally to the language of geology and social history. The Environmental Impact Assessment is already a tool of this methodology and the sub-headings of risk assessment and disaster preparedness need to be woven seamlessly into the process.

There is a need to develop and strengthen local knowledge and history in the educational programmes to make it noticeable and attribute a high profile for its recognition in our daily life. Awareness has to be raised at all levels and the developments of *what-if scenarios* are critical in developing effective urban management. This is in itself an important exercise in as much that it encourages people to have a dream and vision for life. Schools of architecture abound in visionary planning in the aftermath of disasters, business management faculties restructure our society, and children's works on 'drawing their neighbourhood or city' can all be part of our preparedness and the challenges in the face of disasters. This can be facilitated by the use of historic examples which in themselves are less threatening. The recognition of these disasters as *events* as defined in criterion (vi) of the Operational Guidelines to the World Heritage Convention should be encouraged, developing an urban culture through educational and awareness programmes contributing to social resilience.

Through the encouragement of dialogue, in the spirit of the UNESCO constitution, we strive towards peace through mutual assistance and concern. This is the place to foster Cultural Heritage in the lives of the communities being all the more important in the new age of social transformations and urban migrations with the need to re-establish socio-geographic identities. It also highlights our responsibility as architects and designers towards cooperation in the managing of natural disasters in general and those ensuing from climate change in particular. It is the social contract[42] that is critical in this management; it is the mutual responsibilities between communities whether national or local that can generate the acceptance of the other, and reduce conflict by determining the responsibilities of the public, private and NGO sectors to manage the city at times of crisis.

The largest resource of a city is its built fabric and the historic buildings with an added lease of life can often provide conservation solutions using traditional technologies. This brings us back to the need for a comprehensive understanding of preparedness and to ensure its integration into our daily lives as opposed to a separate component dealt with esoterically by experts. As to the accidents of

42 Jean-Jacques Rousseau, *The Social Contract* (Henry Regnery Company 1954, translated by Willmoore Kendall) 2.

life, we need to design in a more proactive way to ensure that a defensible space[43] encompasses all its facets, including those presented by climate change. In short, it should be nothing out of the ordinary.

Acknowledgements

The multiple strands of resilience thinking articulated in this article have been brought together with the help of the colleagues and friends in the European Union 7th Framework security project: 'Designing Safer Urban Spaces' (DESURBS), the UNESCO World Heritage Programme on Disaster Risk Reduction and the debate on climate change and the threats to peace at TU Dresden.

Bibliography

Bandarin, F, 'Foreword' in UNESCO, 'Managing Disaster Risks for World Heritage' World Heritage Resource Manual (June 2010) whc.unesco.org/document/104522 accessed 23 January 2014

von Bertalanffy, L, *General System Theory: Foundations, Development, Applications* (Braziller 1968)

Bogart, D H, 'Feedback, Feedforward, and Feedwithin: Strategic Information in Systems' (1980) 25 Behavioral Science 237

Coaffee, J, 'Rescaling and the Responsibilising of the Politics of Urban Resilience: From National Security to Local Place Making' (2013) 33 Politics 240

Coaffee, J and L Bosher, 'Integrating Counter-terrorist Resilience into Sustainability' (2008) 161 Proceedings of the ICE – Urban Design and Planning 75

Ellenblum, R, *The Collapse of the Eastern Mediterranean: Climate Change and the Decline of the East, 950-1072* (Cambridge University Press 2012)

Evans, J P, 'Resilience, Ecology and Adaption in the Experimental City' (2011) 36 Transactions of the Institute of British Geographers 223

Ewing, R, K Bartholomew, S Winkelman, J Walters, and D Chen, *Growing Cooler: Evidence on Urban Development and Climate Change* (Urban Land Institute 2008)

---

43 Oscar Newman, *Design Guidelines for Creating Defensible Space* (National Institute of Law Enforcement and Criminal Justice, Law Enforcement Assistance Administration, US Dept of Justice 1976) 4–6.

Folke, C, S Carpenter, T Elmqvist, L Gunderson, C S Holling, and B Walker, 'Resilience and Sustainable Development: Building Adaptive Capacity in a World of Transformations' (2002) 31 Journal of the Human Environment 437

Galderisi, A and F F Ferrera, 'Enhancing Urban Resilience in Face of Climate Change: A Methodological Approach' (2012) 5 Journal of Land Use, Mobility and Environment 69

Hamin, E M and N Gurran, 'Urban Form and Climate Change: Balancing Adaptation and Mitigation in the U.S. and Australia' (2009) 33 Habitat International 238

ICOMOS Australia, 'The Burra Charter: The Australia ICOMOS Charter for Places of Cultural Significance 1999' australia.icomos.org/publications/charters/accessed 23 January 2014

ICOMOS, 'The Nara Document on Authenticity' drafted at Nara Conference on Authenticity in Relation to the World Heritage Convention held at Nara, Japan, from 1–6 November 1994 www.icomos.org/charters/nara-e.pdf accessed 23 January 2014

IPCC, 'Climate Change and Water: Technical Paper VI' (June 2008) www.ipcc.ch/pdf/technical-papers/climate-change-water-en.pdf accessed 12 June 2013

IPCC, 'Managing the Risks of Extreme Events and Disasters to Advance Climate Change Adaptation' (2012) www.ipcc-wg2.gov/SREX accessed 12 May 2013

Jigyasu, R and A Vanicka, *Disaster Risk Management of Cultural Heritage in Urban Areas: A Training Guide* (Ritsumeikan University 2013)

Jigyasu, R, M Murhy, G Boccardi, C Marrion, D Douglas, J King, G O'Brian, G Dolcemascolo, Y Kim, P Albrito and M Osihn, 'Heritage and Resilience' 4th Session of the Global Platform for Disaster Risk Reduction (2013) whc.unesco.org/document/122923 accessed 23 January 2014

Maguire, B and P Hagan, 'Disasters and Communities: Understanding Social Resilience' (2007) 22 Australian Journal of Emergency Management 16

McGranahan, G, D Balk and B Anderson, 'The Rising Tide: Assessing the Risks of Climate Change and Human Settlements in Low Elevation Coastal Zones' (2007) 19 Environment and Urbanization 17

Newman, O, *Design Guidelines for Creating Defensible Space* (National Institute of Law Enforcement and Criminal Justice, Law Enforcement Assistance Administration, US Dept of Justice 1976)

O'Brien, G and P Read, 'Future UK Emergency Management: New Wine, Old Skin?' (2005) 14 Disaster Prevention and Management 353

Rose, A, 'Economic Resilience to Natural and Man-made Disasters: Multidisciplinary Origins and Contextual Dimensions' (2007) 7 Environmental Hazards 383

Rousseau, J, *The Social Contract* (Henry Regnery Company 1954, translated by Willmoore Kendall)

UN Habitat, 'City Resilience Profiling Programme: Providing national and local governments with tools for measuring and increading resilience to multi-hazard impacts including those associated with climate change' (UN Habitat information brochure 2012) www.unhabitat.org accessed 23 January 2014

UNESCO, 'Recommendation on the Historic Urban Landscape' (10 November 2011) whc.unesco.org/en/activities/638 accessed 23 January 2014

UNESCO, 'Recommendations of the Second International Workshop on Disaster Risk Reduction to Cultural Heritage Acre (Israel)' (14 November 2009) whc.unesco.org/en/events/610 accessed 23 January 2014

UNISDR, 'Making Cities Resilient' (2012) www.unisdr.org/campaign/resilient cities/accessed 23 January 2014

Vale, L J and T J Campanella, *The Resilient City: How Modern Cities Recover from Disaster* (Oxford University Press 2005)

World Bank, 'Cities and Climate Change: An Urgent Agenda' (2010) 10 Urban Development Series Knowledge Papers 1

Roger-Alexandre Lefèvre*

The Impact of Climate Change on Slow Degradation of Monuments in Contrast to Extreme Events

Abstract In the future, besides extreme events (extreme winds, storms, tornados; extreme precipitations, flooding, flash floods; heat waves, drought; pollution peaks) the Built Cultural Heritage (Monuments) will experience slow events, either in strong correlation with air pollution (recession of façades in limestone or marble; soiling of stone surface; soiling of glass; chemical leaching of mediaeval stained glass; metal corrosion) or in weak correlation with air pollution (salt crystallization in porous walls; freeze-thaw damage in porous materials; submersion of monuments on littoral due to sea level rise; swelling-shrinkage of expansive clay minerals in soils; biomass accumulation on façades in unurban areas).

To assess the impact of slow events, outputs of climate and pollution models are introduced into Dose-Response Functions to allow the projection of the slow weathering of the Built Cultural Heritage on the long-term. In Central and Southern Europe, the conjunction of a slight increase of temperatures and of a local decrease of precipitations, with an important abatement of air pollution, leads to the conclusion that monuments will likely not be threatened by this slow evolution in the 21st century. On the contrary, in Northern Europe, the conjunction of a slight increase of temperatures and an important increase of precipitations with an important abatement of air pollution leads to the conclusion that monuments will be more or less threatened by this slow evolution in the 21st century. This article focuses on some of these slow events.

The goal of this paper is to project and assess the future impact of slow climatic phenomena on Built Cultural Heritage, namely Historical Monuments. This assessment results from the introduction of outputs of climate and pollution models in Dose-Response Function which quantify the behaviour of materials placed in an environment characterized by climate and pollution parameters. At the end, a crucial question is posed: Will monuments be threatened by future slow events in the context of future climate and pollution changes?

* Professor emeritus Paris Est-Créteil University (France) and European University Centre for Cultural Heritage, Ravello (Italy).

I Damages to Built Cultural Heritage Caused by Climate and Pollution

When describing climate and pollution events causing damage to Built Cultural Heritage, it is fundamental to distinguish two types: extreme events and slow events.

Extreme events are short-lived, acute, intensive, recurrent, highly destroying and uncontrollable. They include extreme winds, storms, tornados, extreme precipitations, flooding, flash floods, heat waves, drought and pollution peaks. In the future, both their frequency and intensity are predicted to increase.[1]

Slow events are long-lived, progressive, permanent, less destroying and controllable. Among them, one first category has a strong correlation with air pollution: recession of façades in limestone or marble, soiling of stone surface, soiling of modern glass (haze), chemical leaching of mediaeval stained glass, metal corrosion and biomass accumulation in urban areas. A second category has a weak correlation with air pollution, but hardly depends on geographic location: salt crystallization in porous walls (stone, brick, plaster, frescoes, wall paintings etc), freeze-thaw damage in porous materials, submersion of monuments on littoral due to sea level rise, swelling-shrinkage of expansive clay minerals in soils (affecting the stability of monuments foundations) and biomass accumulation in unurban areas.

In the context of climate change, an inventory of these phenomena and their effects – either extreme or slow – was published by the World Heritage Centre of UNESCO in 2007.[2] However, this list was only descriptive and qualitative, mainly theoretical and at global scale.

Slow phenomena and their outdoor impacts on the materials of Built Heritage were quantified and mapped at the European continental scale in the "Noah's Ark" European Project (2004–2007).[3] Although these slow phenomena have not been mapped, they were quantified at a local scale in London for recession and

---

1 IPCC, *Fourth assessment report: climate change* (Cambridge University Press 2007).

2 UNESCO, 'Climate Change and World Heritage. Report on Predicting and Managing the Impacts of Climate Change on World Heritage and Strategy to Assist States Parties to Implement Appropriate Management Responses' World Heritage Reports 22 (1 May 2007).

3 Alessandra Bonazza et al, 'Mapping the impact of climate change on surface recession of carbonate buildings in Europe' (2009) 407 Science of the Total Environment 2039; Cristina Sabbioni, Peter Brimblecombe and May Cassar (eds), *The Atlas of Climate Change Impact on European Cultural Heritage* (Anthem Press 2010).

soiling of stone;[4] in Oviedo, Prague and Paris only for recession of stone;[5] and in Paris for damage on glass and stained glass.[6] Maps of the effects of climate and pollution on limestone at the scale of two different metropolises, Istanbul[7] and Madrid,[8] by means of the Dose-Response Functions, have been recently published. However, the Istanbul study concerns only the current period while the one on Madrid gives projections until 2025.

Indoor effects on Built Cultural Heritage will be quantified and mapped at the European scale and at the scale of individual buildings in the frame of "Climate for Culture" European Project (2010–2014).[9]

The Council of Europe in Strasbourg pays attention to the impacts of climate change on materials of Built Cultural Heritage by producing reports[10] and funding activities in this field at the European University Centre for Cultural Heritage, Ravello (Italy): doctoral courses[11] and international workshops.[12]

II The Method for Predicting and Assessing the Slow Impacts of Climate Change on Built Cultural Heritage

This method consists of introducing the outputs of climate and pollution models into the Dose-Response Functions (DRF). Climate models (e.g. English

4 Peter Brimblecombe and Carlota M Grossi, 'Millennium-long Recession of Limestone Façades in London' (2008) 56 Environmental Geology 463; Peter Brimblecombe and Carlota M Grossi, 'Millennium-long Damage to Building Materials in London' (2009) 407 Science of the Total Environment 1354.

5 Carlota M Grossi et al, 'Predicting Twenty-first Century Recession of Architectural Limestone in European Cities' (2008) 56 Environmental Geology 455.

6 Anda Ionescu et al, 'Long-term Damage to Glass in Paris in a Changing Environment' (2012) 431 Science of the Total Environment 151.

7 Ferhat Karaca, 'Mapping the Corrosion Impact of Air Pollution on the Historical Peninsula of Istanbul' (2013) 14 Journal of Cultural Heritage 129.

8 Daniel De la Fuente et al, 'Mapping Air Pollution Effects on Atmospheric Degradation of Cultural Heritage' (2013) 14 Journal of Cultural Heritage 138.

9 Johanna Leissner, 'The Impact of Climate Change on Historic Buildings and Cultural Property' (2011) 3 UNESCO Today 44; Constanze Fuhrmann and Johanna Leissner, 'Climate for Culture Project: First Results' (2012) www.climateforculture.eu/index.php?inhalt=dissemination.publications accessed 24 January 2014.

10 Cristina Sabbioni et al, 'Vulnerability of Cultural Heritage to Climate Change' Report to Council of Europe (20 November 2008) AP/CAT (2008) 44.

11 Roger A Lefèvre and Cristina Sabbioni, 'Vulnerability of Cultural Heritage to Climate Change' (European Master-Doctorate Course, Council of Europe, Strasbourg 2009).

12 Roger A Lefèvre and Cristina Sabbioni (eds), *Cultural Heritage and Climate Change. Proceedings of the Ravello International Workshop 2009* (Bari: Edipuglia 2009).

HADLEY, French ARPEGE-CLIMAT, European ENSEMBLES) propose: future temperature (T), air relative humidity (RH) and amount and acidity (H^+) of rain. Pollution models (e.g. European GAINS) propose future concentrations of SO_2, HNO_3, O_3 and particulate matter (PM).

The *DRFs* are equations allowing the projection of the behaviour of a material in new environmental and climatic conditions, current or future, or even reconstructing its past behaviour.[13] The *doses* are the parameters influencing the response; in this case, they correspond to climatic (meteorological) and environmental (pollution) factors. Their influence is different according to the exposure conditions: dry (sheltered from rain) or wet (unsheltered). The *response* is a measurable modification of the material: corrosion, recession, soiling, leaching etc.

The *general expression* of a DRF is:

$$\text{Response (damage)} = \text{Dose}_{dry}\,(T, RH, [SO_2], [HNO_3], [O_3], [PM], t\ldots) + \text{Dose}_{wet}\,(\text{Rain}, [H^+], t\ldots)$$

The DRFs are obtained either by exposure in *atmospheric simulation chambers* where the different doses are simulated, or, more frequently, by *exposure campaigns on real sites* with various environmental and climatic parameters, which are measured around the experience. The exposure duration 't' is accounted as a dose. Only the measured doses are taken into account, excluding all others, even if they should be included as possible causes of the response, but unknown at the time of the experience.

These functions present a crucial interest as they can be mapped in the same way as climatic and pollution factors.

III Recession of Façades in Limestone or Marble

When unsheltered, the parts of the façades in limestone or marble exposed to rain erode by dissolution of carbonates and their surface begins to recede. A damage function (DRF) was established by Lipfert[14] *after laboratory experiments* on the dissolution of calcite ($CaCO_3$), the main component of limestone and marble:

$$\text{Recession}_{\mu m\ y^{-1}} = 18.8\ \text{Rain}_{m\ y^{-1}} + 0.016\ [H^+]_{\mu mol\ l^{-1}}\ \text{Rain}_{m\ y^{-1}} + 0.18\,(V_{dS\ cm\ s^{-1}}\,[SO_2]_{\mu g\ m^{-3}} + V_{dN\ cm\ s^{-1}}\,[HNO_3]_{\mu g\ m^{-3}})$$

13 Vladimir Kucera et al, 'UN/ECE ICP materials dose-response functions for the multipollutant situation' (2007) 7 Water Air and Soil Pollution Focus 249.

14 Frederick W Lipfert, 'Atmospheric damage to calcareous stones: comparison and reconciliation of recent experimental findings' (1989) 23 Atmospheric Environment 415.

If measurements of HNO_3 are not available, Kucera[15] suggests the following conversion formula:

$$[HNO_3]_{\mu g\ m^{-3}} = 516e^{-3400/(T+273)}\ ([NO_2]_{\mu g\ m^{-3}}\ [O_3]_{\mu g\ m^{-3}}\ RH_{\%})^{0.5}$$

The first factor of the Lipfert equation represents the action of clean rain (karst effect), the second the action of acid rain and the third the action of acid dry deposition (V_{dS} and V_{dN} being respectively the deposition velocity of SO_2 and HNO_3). Therefore this DRF underlines that the recession of façades in limestone or marble depends both on climate (clean rain) and pollution (wet/dry pollution: acid rain, SO_2, HNO_3).

This DRF, established by Lipfert, was applied to the reconstruction of the past and the projection of the future recession rates of porous Portland limestone, in London from 1100 to 2100[16] (*Figure 1*). The karst effect dominates the earliest period and continues, only with subtle changes, throughout. Pollution damage to stone increases rapidly in the late 1600s as the result of the dramatic increase of the use of high-sulphur coal. This use ultimately dropped at the end of 20th century.

Figure 1: Recession rates of porous Portland limestone in London 1100–2100. The contributions to recession rates of acid rain, dry deposition and karst effect are added to show the sum of the contributions (After Brimblecombe & Grossi, 2008).

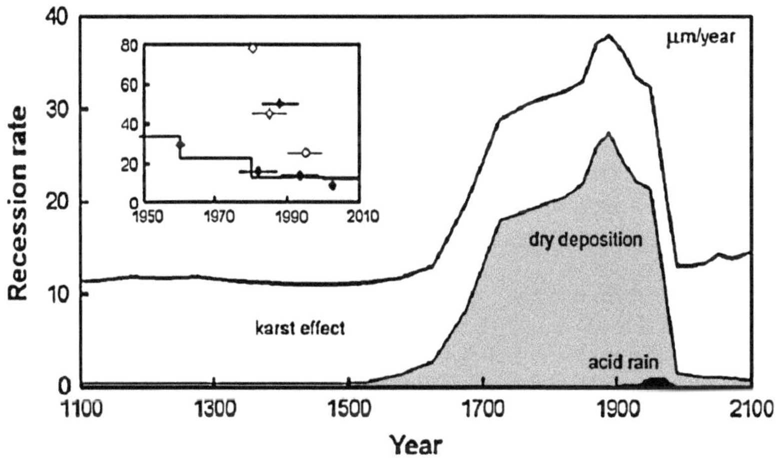

15 Vladimir Kucera, 'Model for multipollutant impact and assessment of threshold levels for Cultural Heritage' (Swedish Corrosion Institute, Stockholm 2005) www.corrinstitute.se/multi-assess/web/page.aspx accessed 24 January 2014.

16 Brimblecombe and Grossi (n 4).

Figure 1 indicates clearly that the recession rates of Portland limestone in *London* for the period beginning in 2000 are of 12-14 μm y^{-1}. These rates are regularly, but slowly, increasing throughout the 21st century due to the increasing of the karst effect, a consequence of increasing temperature and atmospheric concentrations of CO_2, although rainfall may decline.[17]

A second DRF is given by Tidblad *et al.*[18] after *field campaigns of exposure of samples of Portland limestone* to various real current environments:

$$\text{Recession}_{\mu m} = 2.7\,[SO_2]^{0.48}\mu g\ m^{-3}\exp(-0.018T)\,t_y^{0.96} + 0{,}019\ \text{Rain}_{mm}\,[H+]_{\mu mol\ l^{-1}}\,t_y^{0.96}$$

Measurements of the recession rates of Portland limestone were performed at the St Paul Cathedral in London for three decades (1980–2010).[19] The results of these measurements were compared to the calculations based on the DRF for *calcite* by Lipfert[20] and for the same *limestone* by Tidblad *et al*[21] (*Table 1*):

Table 1: Annual recession rates (in μm) measured and calculated by decades of Portland limestone at St Paul Cathedral in London[22]

Period	**Measures at St Paul Cathedral**	**DRF Lipfert (1989)**	**DRF Tidblad *et al.* (2001)**
1980–1990	49	14.72	8.54
1990–2000	38	11.78	5.18
2000–2010	35	12.64	3.74

There is an important difference between the results obtained from the two DRFs and between them and the results of measurements on the cathedral: Tidblad's DRF underestimates Lipfert's DRF (results are lower from the half to the third). All of these calculations are significantly lower than the in the field

17 Bonazza et al (n 3).

18 Johan Tidblad et al, 'UN ECE ICP Materials: Dose-response Functions on Dry and Wet Acid Deposition Effects over 8 Years of Exposure' (2001) 130 Water, Air, Soil Pollution 1457.

19 Rob Inkpen et al, 'Modelling the Impact of Changing Pollution Levels on Limestone Erosion Rates in Central London, 1980–2010' (2012) 61 Atmospheric Environment 476; Rob Inkpen et al, 'Thirty Years of Erosion and Declining Atmospheric Pollution at St Paul's Cathedral, London' (2012) 62 Atmospheric Environment 521.

20 Lipfert (n 14).

21 Tidblad et al (n 18).

22 Inkpen et al (n 19).

measurements (from the tenth to the third). Nevertheless all of these results are in the same magnitude. Inkpen *et al* wrote:

> "Researchers using dose-response functions are well aware that there are problems with their construction and use. Likewise, researchers who use Micro Erosion Meter or stone tablets to monitor erosion are well aware that this method is not without […] problems".[23]

According to the results of the Noah's Ark European Project, *at European scale,* the recession of façades in limestone or marble will remain constant in the *near future* (2013–2039) but, in the *far future* (2070–2099), although it will remain constant in Central Europe, it will hardly decrease in Southern Europe and it will increase in Northern Europe.

IV Soiling (Blackening) of Façades in Limestone or Marble

Contrary to the previous location, the parts of the façades in limestone or marble sheltered from rain accumulate particles (Particulate Matter, PM) not washed by rain after their dry deposition. An equation, having the value of DRF, links the reflectivity or lightness R_t after time t to the initial reflectance R_0 (clean stone) and the concentration of fine particulate matter (PM_{10}) (<10 μm) with a constant rate for blackening k_s[24]:

$$R_t \% = R_0 \% \exp(-k_s [PM_{10}]_{\mu g\ m^{-3}}\ t_y)$$

This DRF assumes that soiling of façades in limestone or marble depends only on pollution (PM_{10}).

Figure 2 shows the evolution over the last millennium of the blackening of a new sample of Portland limestone in *London.* As for recession (*Figure 1*), the blackening increased from the 1600s with a peak at the end of the 19th century and decreased in the 20th century. This decrease will continue throughout the 21st century. The shaded area points out the contribution of busy streets where the PM_{10} is taken as 50 per cent higher than urban background and the particles a few per cent darker.

23 Inkpen et al (n 19).

24 Brimblecombe and Grossi (n 4).

Figure 2: Blackening rate (% y^{-1}) as loss of reflectance of new sample of Portland limestone (80 % initial reflectance) exposed in London from 1100 to 2100. EC/µg m^{-3}: Elemental Carbon concentration; % EC: percentage of Elemental Carbon with respect to PM_{10}. The shaded area shows the increase of PM_{10} in busy streets (After Brimblecombe & Grossi 2009)

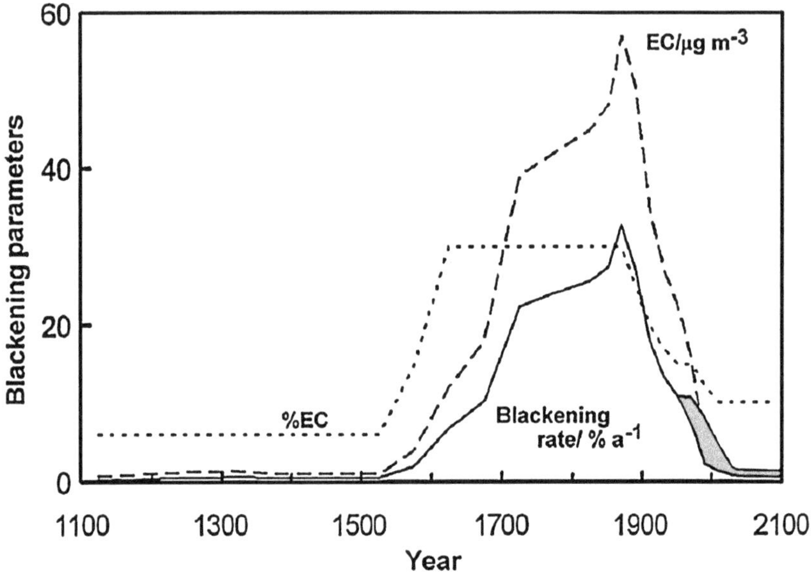

Although the quantification and the mapping of soiling of façades were not achieved in the Noah's Ark Project, *at the European scale,* there is a good reason to do it. Indeed, soiling is a complementary phenomenon of recession (soiling in sheltered zones and recession in unsheltered zones of the same façade), the two phenomena can be quantified and mapped at the same European scale and their long-term evolution might be estimated as well.

V Soiling of Modern Si-Ca-Na Glass (Haze)

Modern Si-Ca-Na glass is used for windows and building façades as well as for double protective glazing of stained glass windows. Due to its chemical composition, this modern glass is highly durable. From the physical point of view, haze is mainly caused by particle deposition and the development of microcrystals, mainly sulphates (neocrystallisations), on the glass surface. Glass soiling can be accessed through an optical measurement: the haze, defined as the ratio between the diffuse transmitted light and the direct transmitted light, expressed in a percentage. The corresponding DRF is expressed as[25]:

$$\text{Haze}_{\%} = (0.2529\,[SO_2]_{\mu g\ m^{-3}} + 0.108\,[NO_2]_{\mu g\ m^{-3}} + 0.1437\,[PM_{10}]_{\mu g\ m^{-3}}).\ 1 / (1 + (382 / t_{days})^{1.86})$$

One can notice from the previous expression that Haze of modern glass seems to depend only on pollution (SO_2, NO_2, PM_{10}), the climate parameters having no significant contribution.

Figure 3 shows the evolution from 1500 to 2100 of haze development, from the first year of exposure of a pristine sample of Si-Ca-Na glass in *Paris*. Comparing *Figures 1, 2 and 3*, the picture emerging in London is broadly similar to Paris, however, the period of rapid damage to materials is longer and begins earlier in London. This difference reflects the much earlier adoption of coal as a fuel in London.[26] A general abatement is projected in the 21st century.

Figure 3: Estimated haze in % development during the first year of exposure in Paris of modern Si-Ca-Na glass using the DRF including (dotted line) and excluding (solid line) the effect of NO_2 (After Ionescu et al 2012)

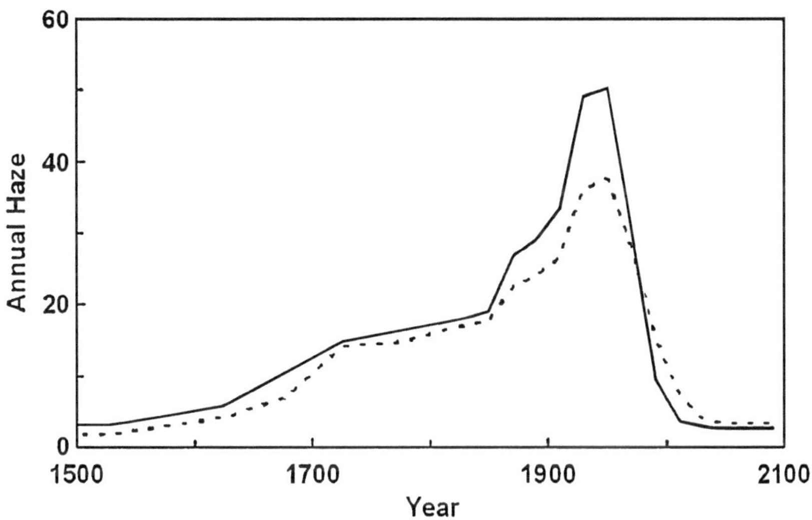

25 Tiziana Lombardo et al, 'Dose–response Function for the Soiling of Silica–soda–lime Glass due to Dry Deposition' (2010) 408 Science of the Total Environment 976.

26 Ionescu et al (n 6).

Modern glass was not a material studied in the Noah's Ark Project. For this reason its soiling (haze) was not mapped and not considered on the long term at the European scale. Nevertheless, its development depending only on pollution parameters it is possible predicting its general abatement all over Europe during the 21st century.

VI Chemical Leaching of Mediaeval Si-Ca-K Stained Glass

In the goal of establishing a DRF for the weathering of mediaeval Si-Ca-K stained glass, less durable than Si-Ca-Na modern glass, Melcher & Schreiner[27] measured the leaching depth of the ions K^+ and $Ca2^+$ in mediaeval-like model glass specimens exposed to the environment of several sites *sheltered from rain*. The superficial leaching of K and Ca leads to the formation of a hydrated silica gel layer at the surface of stained glass, which plays a relatively protecting role.

DRFs, as a result of regression calculations for the leaching depths of the two main alkali ions d(K) and d(Ca) in the glass, are given by the authors:

$$d(K)_{\mu m} = -0.64 + (0.03RH\% + 0.04\,[SO_2]_{\mu g\ m^{-3}})\sqrt{t}_{y} - (0.05T_{°C} + 2.03.\ 1/[NO_2]_{\mu g\ m^{-3}})\ t_{y}$$

$$d(Ca)_{\mu m} = -0.79 + (0.03RH\% + 0.03\,[SO_2]_{\mu g\ m^{-3}})\sqrt{t}_{y} - (0.04T_{°C} + 1.91.\ 1/[NO_2]_{\mu g\ m^{-3}})\ t_{y}$$

In conclusion, DRFs established for the Si-Ca-K glass reveal that leaching of mediaeval glass *sheltered from rain* depends both on pollution (SO_2, NO_2) and climate (RH, T) parameters.

An application of these two DRFs is illustrated in *Figure 4*: the trends in the depth to which the annual loss occurs for both K and Ca in stained glass *sheltered from rain* in *Paris* are seen to reflect changes in pollutant concentrations over time. The decrease observed in the late 20th century will not attain very low values in the 21st century due to climate factors (RH, T).

27 Michael Melcher and Manfred Schreiner, 'Quantification of the influence of atmospheric pollution on the weathering of low-durability potash-lime-silica glasses' (2007) 49 Pollution atmosphérique 13; Michael Melcher and Manfred Schreiner, 'Impact of Climate Change on Medieval Stained Glass' in Roger A Lefèvre and Cristina Sabbioni (eds), *Cultural Heritage and Climate Change* (Edipuglia 2010) 93–99.

Figure 4: The depth of damage in μm under the surface of a stained glass window with respect to leaching of potassium (dotted line) and calcium (solid line) over the first year of exposure in Paris sheltered from rain (After Ionescu et al 2012)

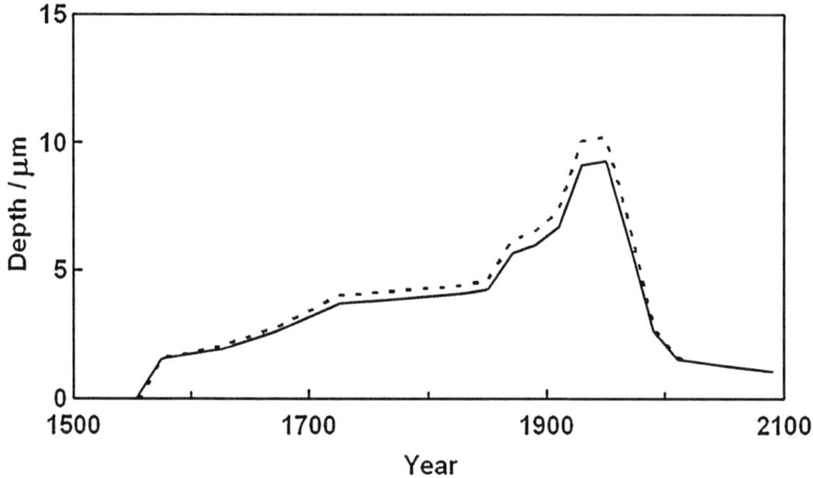

Si-Ca-K mediaeval-like model glass samples were exposed *unsheltered from rain* to the Parisian atmosphere on the Northern tower of the Saint Eustache Church, in the pedestrian quarter 'Les Halles' in the centre of *Paris*.[28] The thickness of the leached layer is given by the author as a function of the duration of exposure in months (*Table 2*).

Table 2: Thickness in μm of the superficial leached layer of Si-Ca-K glass exposed to rain in Paris unsheltered from rain resulting from measurements on Scanning Electron Microscopy Images and Electron Probe Microanalysis (6 and 12 Month) (After Gentaz, 2011, Tabl. 28, p. 153)

		6 months	12 months	15 months	24 months	36 months	48 months
Thickness (μm)	Min.	–	–	–	1.6	0.6	0.6
	Mean	2.5	4.1	–	9.3	13.8	17.0
	Max.	–	–	11	17.5	25.7	37.0

28 Lucile Gentaz, *Simulation et modélisation de l'altération des verres de composition médiévale dans l'atmosphère urbaine* (Thèse de doctorat University Paris Est-Créteil 2011).

Other Si-Ca-K mediaeval-like glass model samples were *exposed to rain* at the Sainte Chapelle in *Paris* for 12 months during the European Project VIDRIO (2001-2004).[29] The thickness of the leached layer was 0.6 µm, that is less than the thickness (4.1 µm) measured by Gentaz[30] for the same duration *exposed to rain*, less than the results of the calculations performed for the same duration in Paris by Ionescu *et al*[31] (*Figure 4*) by means of the two DRFs given by Melcher & Schreiner[32] including the outputs of HADLEY and GAINS models (1 to 2 µm in average), but in *sheltered situation.* Although being quite different, these results are in the same range both for the sheltered and unsheltered situations or those resulting from calculations (DRFs and climate and pollution models).

Contrary to modern glass, mediaeval stained glass was studied in the Noah's Ark Project[33] on the basis of the DRFs established by Melcher & Schreiner.[34] The conclusions were that a minor decrease in stained glass leaching all over Europe is predicted, but meaning that mediaeval glass will remain at risk in many areas.

The four diagrams (*Figures 1, 2, 3 and 4*) show clearly that in the 21st century, at least in London and Paris, the general decrease of two slow phenomena are linked exclusively to pollution: soiling of stone and haze of modern glass. They also show the different projected evolutions of two slow phenomena linked to both climate and pollution: erosion-recession of stone will slightly increase and leaching of mediaeval stained glass will slightly decrease.

VII Corrosion of Metals

For two metals, steel and zinc, the following DRFs were established[35]:

$$\mathrm{Corr}_{\mathrm{steel}} = 1.58\,[\mathrm{SO_2}]^{0.52}_{\mu g\ m^{-3}}\, e^{0.20\,\mathrm{RH\%} + f_{\mathrm{steel}}(T_{°C})} + 0.102\, D_{\mathrm{Cl}}\, e^{0.033\mathrm{RH\%} + 0.040 T_{°C}}$$

$$\mathrm{Corr}_{\mathrm{zinc}} = 0.011\,[\mathrm{SO_2}]^{0.44}_{\mu g\ m^{-3}}\, e^{0.046\,\mathrm{RH\%} + f_{\mathrm{zinc}}(T_{°C})} + 0.0175\, D_{\mathrm{Cl}}\, e^{0.008\mathrm{RH\%} + 0.085 T_{°C}}$$

29 Adriana Bernardi et al, 'Conservation of stained glass windows with protective glazing: Main results from the European VIDRIO research programme' (2013) 14 Journal of Cultural Heritage 527.

30 Gentaz (n 28).

31 Ionescu et al (n 6).

32 Melcher and Schreiner (n 27).

33 Sabbioni et al (n 3).

34 Melcher and Schreiner (n 27).

35 Kucera et al (n 13); Johan Tidblad, 'Atmospheric corrosion of metals in 2010–2039 and 2070–2099' (2012) 55 Atmospheric Environment 1.

This assumes that corrosion of metals depends both on climate (RH, T) and pollution (SO_2, Chloride).

The expected evolution during the 21st century was mapped in the frame of the European Noah's Ark Project[36] (*Figure 5*).

Figure 5: Map of the expected corrosion risk for metals at the end of 21st century (2070–2099) issued by the European Noah's Ark Project (After Sabbioni et al 2010)

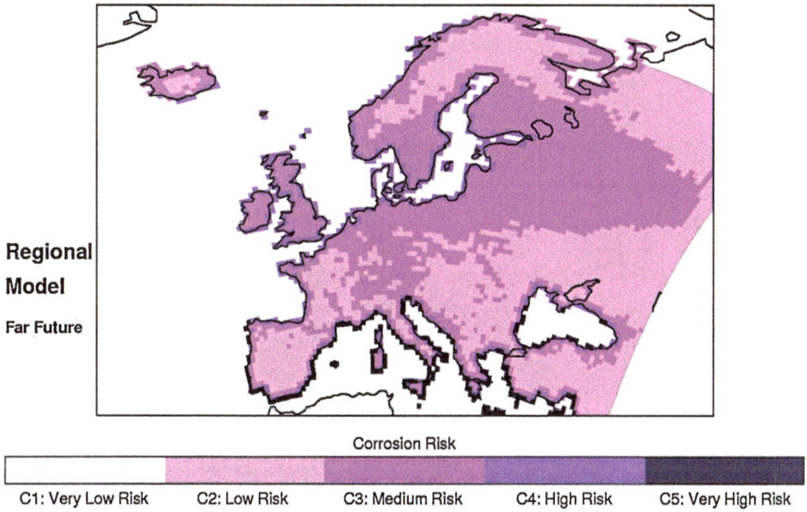

The atmospheric corrosion of metals in inland areas is expected to increase in the Northern Europe (where the trend will be dominated by the effect of average annual temperature, as a maximum corrosion effect is observed at 10°C) and to decrease in Southern Europe. In European coastal areas affected by high chloride deposition, corrosion is expected to increase.[37]

VIII Biomass Accumulation in Unurban Areas

Organism growth on monuments is of particular interest, both as an agent of aesthetic change as well as its possible role in contributing to stone deterioration or protecting against deterioration. Biomass production was correlated with climate conditions, the main parameters defining bioclimate being precipitation

36 Sabbioni et al (n 3); Tidblad (n 35).
37 Sabbioni et al (n 3).

and temperature. Changes in these parameters affect organism activity and hence changes in the biomass.

The accumulation of biomass (Lichens, Algae, Mosses, Fungi etc) on façades *in unurban areas* is linked to rain and temperature, depending only on climate parameters[38]:

$$B_{mg\ cm^{-2}} = e^{(-0.964 + 0.003\ Rain_{mm\ y^{-1}} - 0.01 T_{°C})}$$

According to this DRF, rain has a positive effect while temperature has a negative effect. It has to be taken into consideration that in urban areas pollution has a negative effect on biomass growth.

Figure 6: Biomass accumulation projection for 2070–2099 (After Gómez-Bolea et al 2012)

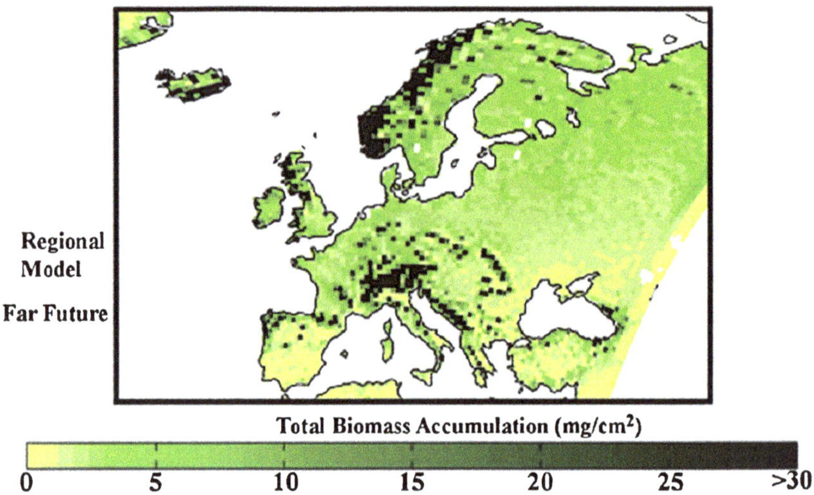

The models on climate change predict an increase in temperature and precipitation in Northern areas of Europe for the far future (2070–2099), which would lead to a higher accumulation of biomass. Contrary to the North, a significant reduction in precipitation is expected in Southern areas of Europe, associated with a lower biomass accumulation in such areas.[39]

38 Antonia Gómez-Bolea et al, 'Mapping the Impact of Climate Change on Biomass Accu- mulation on Stone' (2012) 13 Journal of Cultural Heritage 254.

39 Gómez-Bolea (n 38).

IX Will Monuments Really be Threatened by Slow Events Related to Future Climate Change and Future Pollution?

In Northern Europe, the conjunction of *a slight increase of temperatures*, *an important increase of precipitations* and *an important abatement of air pollution* leads to the conclusion that monuments will be more or less threatened by this *slow* evolution during the 21st century.

On the contrary, in Central and Southern Europe, the conjunction of *an increase of temperatures, a decrease of precipitations* and *an important abatement of air pollution* lead to the conclusion that monuments are not likely to be threatened by this *slow* evolution during the 21st century.

Nevertheless, *climate change is obviously a reality* demonstrated by observations, measurements and modelling. In addition, besides slow events, *extreme events* are serious threats for monuments. This is because they will likely be reinforced both by frequency and strength in the future.

The basic research in the field of Cultural Heritage is the scientific determination of the climate parameters that are most critical to heritage building's materials, named *Heritage climatology* by Brimblecombe.[40] There are a number of climate models and they all need to be tested and validated for the foreseen impact on Cultural Heritage. There are particular problems in applying existing models as the right spatial and temporal scales are required. Although the spatial resolution of models has improved in recent years, this is not always true of temporal scales as heritage research not only needs data on the long term, but often at annual or daily time intervals. It is essential to improve and test Dose-Response Functions for a wide range of heritage materials for past, present and future climates in indoor and outdoor environments.[41]

Bibliography

Bernardi, A, F Becherini, M Verità, P Ausset, M Bellio, U Brinkmann, H Cachier, A Chabas, F Deutsch, M P Etcheverry, F Geotti Bianchini, R H M Godoi, V Kontozova-Deutschj, R A Lefèvre, T Lombardo, P Mottner, C Chiara Nicola, I Pallot-Frossard, S Rölleke, H Römich, G Sommariva, M Vallotto and R Van Grieken, 'Conservation of stained glass windows with protective glazing: Main results from the European VIDRIO research programme' (2012) 14 Journal of Cultural Heritage 527

40 Peter Brimblecombe, 'Heritage Climatology' in Roger A Lefèvre and Cristina Sabbioni (eds), *Cultural Heritage and Climate Change* (Edipuglia 2010) 49–63.

41 Sabbioni et al (n 3).

Bonazza, A, P Messina, C Sabbioni, C M Grossi and P Brimblecombe, 'Mapping the Impact of Climate Change on Surface Recession of Carbonate Buildings in Europe' (2009) 407 Science of the Total Environment 2039

Brimblecombe, P, 'Heritage Climatology' in Lefèvre R A and C Sabbioni (eds), *Climate Change and Cultural Heritage* (Edipuglia 2010)

Brimblecombe, P and C M Grossi, 'Millennium-long Recession of Limestone Façades in London' (2008) 56 Environmental Geology 463

Brimblecombe, P and C M Grossi, 'Millennium-long Damage to Building Materials in London' (2009) 407 Science of the Total Environment 1354

De la Fuente, D, J M Vega, F Viejo, I Diaz and M Morcillo, 'Mapping Air Pollution Effects on Atmospheric Degradation of Cultural Heritage' (2013) 14 Journal of Cultural Heritage 138

Fuhrmann, C and J Leissner, 'Climate for Culture project: first results' (2012) www.climateforculture.eu/index.php?inhalt=dissemination.publications accessed 24 January 2014

Gentaz, L, *Simulation et modélisation de l'altération des verres de composition médiévale dans l'atmosphère urbaine* (Doctorate Thesis University Paris Est-Créteil 2011)

Gómez-Bolea, A, E Llop, X Arino, C Saiz-Jimenez, A Bonazza, P Messina and C Sabbioni, 'Mapping the Impact of Climate Change on Biomass Accumulation on Stone' (2012) 13 Journal of Cultural Heritage 254

Grossi, C M, A Bonazza, P Brimblecombe, I Harris and C Sabbioni, 'Predicting Twenty-first Century Recession of Architectural Limestone in European Cities' (2008) 56 Environmental Geology 455

Inkpen, R, H Viles, C Moses and B Baily, 'Modelling the Impact of Changing Pollution Levels on Limestone Erosion Rates in Central London, 1980–2010' (2012) 61 Atmospheric Environment 476

Inkpen, R, H Viles, C Moses, B Baily, P Collier, S T Trudgill and R U Cooke, 'Thirty years of Erosion and Declining Atmospheric Pollution at St Paul's Cathedral, London' (2012) 62 Atmospheric Environment 521

Ionescu, A, R A Lefèvre, P Brimblecombe and C M Grossi, 'Long-term Damage to Glass in Paris in a Changing Environment' (2012) 431 Science of the Total Environment 151

IPCC, *Fourth Assessment Report: Climate Change* (Cambridge University Press 2007)

Karaca, F, 'Mapping the Corrosion Impact of Air Pollution on the Historical Peninsula of Istanbul' (2013) 14 Journal of Cultural Heritage 129

Kucera, V, 'Model for Multipollutant Impact and Assessment of Threshold Levels for Cultural Heritage' (Swedish Corrosion Institute Stockholm 2005) www.corr-institute.se/multiassess/web/page.aspx accessed 24 January 2014)

Kucera, V, J Tidblad, K Kreislova, D Knotkova, M Faller, D Reiss, D Snethlage, T Yates, J Henriksen, M Schreiner, M Melcher, M Ferm, R A Lefèvre and J Kobus, 'UN/ECE ICP Materials Dose-response Functions for the Multipollutant Situation' (2007) 7 Water Air and Soil Pollution Focus 249

Lefèvre, R A and C Sabbioni (eds), *Cultural Heritage and Climate Change. Proceedings of the Ravello International Workshop 2009* (Bari: Edipuglia 2009)

Lefèvre, R A and C Sabbioni, *Vulnerability of Cultural Heritage to Climate Change* (European Master-Doctorate Course, Council of Europe Strasbourg 2009)

Leissner, J, 'The Impact of Climate Change on Historic Buildings and Cultural Property' (2011) 3 UNESCO Today 44

Lipfert, F W, 'Atmospheric Damage to Calcareous Stones: Comparison and Reconciliation of Recent Experimental Findings' (1989) 23 Atmospheric Environment 415

Lombardo, T, A Ionescu, A Chabas, R A Lefèvre, P Ausset and Y Candau, 'Dose–response Function for the Soiling of Silica–soda–lime Glass due to Dry Deposition' (2010) 408 Science of the Total Environment 976

Melcher, M and M Schreiner, 'Quantification of the Influence of Atmospheric Pollution on the Weathering of Low-durability Potash-lime-silica Glasses' (2007) 49 Pollution atmosphérique 13

Melcher, M and M Schreiner, 'Impact of Climate Change on Medieval Stained Glass' in Lefèvre R A and C Sabbioni (eds), *Cultural Heritage and Climate Change* (Edipuglia 2010)

Sabbioni, C, M Cassar, P Brimblecombe and R A Lefèvre, 'Vulnerability of Cultural Heritage to Climate Change' Report to Council of Europe (20 November 2008) AP/CAT (2008) 44

Sabbioni, C, P Brimblecombe and M Cassar (eds), *The Atlas of Climate Change Impact on European Cultural Heritage* (Anthem Press 2010)

Tidblad, J, 'Atmospheric Corrosion of Metals in 2010–2039 and 2070–2099' (2012) 55 Atmospheric Environment 1

Tidblad, J, V Kucera, A A Mikhailov, J Henriksen, K Kreislova and T Yates, 'UN ECE ICP Materials: Dose-response Functions on Dry and Wet Acid Deposition Effects over 8 Years of Exposure' (2001) 130 Water, Air, Soil Pollution 1457

UNESCO, 'Climate Change and World Heritage. Report on Predicting and Managing the Impacts of Climate Change on World Heritage and Strategy to Assist States Parties to Implement Appropriate Management Responses' World Heritage Reports 22 (1 May 2007)

Reza Ardakanian & Stephan Hülsmann*

Impact of Global Change on World Heritage and on Environmental Resources: The Need for an Integrated Management Approach

Abstract Natural Heritage sites are obviously closely related to and affected by environmental resources (mainly water and soil) of the area they are situated in. However, also Cultural Heritage sites are not only affected by e.g. water (via precipitation, water level etc), but in most cases are part of a cultural landscape in which the surrounding nature and its environmental resources are a critical element. This notion certainly holds for Cultural Heritage in which water is an essential part. Further, any change in land use of heritage sites, be it climate related or man-made, may have a strong impact. Thus, virtually all Cultural Heritage Sites cannot be managed and maintained without considering their natural environment. Climate change directly affects both Natural and Cultural Heritage sites (e.g. changing temperature and precipitation patterns and changing, often increasing frequency of catastrophic events), but perhaps more importantly, it has an indirect effect via its impact on the environmental resources. In addition, climate change may increase the pressure on heritage sites by increasing human impact, since the people in the region may face a shortage of natural resources and/or increase in population density or move to urban areas. This means one has to consider impacts of global change rather than only climate.

Mitigating the direct and indirect impacts of global change on heritage sites and/or adapting to it requires an integrated approach. Protection and management of heritage sites must be considered in the context of managing environmental

* Prof. Reza Ardakanian is founding director of the United Nations University Institute for Integrated Management of Material Fluxes and of Resources (UNU-FLORES), Dresden, Germany. Experiencing boards of various international programs or organizations, such like UNESCO-IHP, UNESCO-IHE, the International Hydropower Association and UNU-EHS, he has served as director of the UN-Water Decade Programme on Capacity Development hosted by UNU since 2007 and as vice-rector *ad interim* of UNU in Europe.
Dr. Stephan Hülsmann holds the position of academic officer of UNU-FLORES and specialist in Systems and Flux Analysis considering Global Change Assessment.

resources in the region in a sustainable way. In addition, especially in developing countries, these management strategies need to be embedded in a suite of measures and activities in capacity development addressing all relevant stakeholders. The mission of the United Nations University Institute for Integrated Management of Material Fluxes and of Resources (UNU-FLORES) is to address the urgent need for integrated and sustainable management strategies for environmental resources, focusing on water, soil and waste. Such an integrated approach to the management of inter-related resources may also be termed a nexus approach to the management of water, soil and waste. Implementing such strategies, considering capacity development both at the individual and institutional level will be one means to manage and maintain World Heritage, be it natural or cultural.

I Heritage Sites and Their Environment

The UNESCO World Heritage Center currently lists 119 *water-related* properties (lakes) of outstanding universal value, among them 59 listed as Cultural Heritage, 51 as Natural Heritage and 8 as mixed sites.[1] Natural heritage sites, often representing national parks and reserves, are protected mainly as areas of exceptional natural beauty and aesthetic importance, as outstanding examples representing major stages of earth's history, or significant ecological and biological processes in the evolution of ecosystems or based on their biological diversity. Usually, human impact and use of natural resources in these areas is prohibited or very restricted. For Natural Heritage sites it is clear and obvious that they are strongly related and interlinked with their natural environment. Any change in land-use, increasing usage or withdrawal of water, mostly associated with pollution would certainly directly affect the Natural Heritage site.

Is this close relation and dependency of Heritage sites on the natural environment and the surrounding environmental resources indeed so specific to Natural Heritage? In fact, we are making the case here that it applies also to Cultural Heritage sites in a similar way. When talking about specific monuments and buildings, the relation is obviously not as close as to Natural Heritage sites, although they are still affected by environmental resources, e.g. by water *via* precipitation (potentially polluted), groundwater level etc. However, the relation of Cultural Heritage sites to the surrounding environmental resources is in many (if not in

1 As of November 2014; for dynamic reference please see whc.unesco.org/en/list inserting *lake* as search criteria.

most) cases much closer and rather similar to Natural Heritage sites: Cultural Heritage sites are typically closely embedded into their natural environment with its natural resources; they are part of a landscape (if not the landscape itself is the Cultural Heritage). This means, also in the case of Cultural Heritage, any change in land-use, be it climate-related or man-made, or any change in the availability or quality of water resources, will have an impact.

Some arbitrarily chosen prominent examples to showcase the close interrelatedness with environmental resources are briefly described as follows (*Figures 1, 2, 3 and 4*); please note that all are explicitly classified as Cultural Heritage:

- West Lake Cultural Landscape of Hangzhou (China):[2] This is considered an outstanding example of a cultural landscape where natural elements, farmed landscape and artificial elements manifest a *perfect fusion*;
- Angkor (Cambodia):[3] This site includes a concentration of impressive monuments, "closely linked to their geographic context", including (artificial) water bodies, forests and cultivated land;
- Shushtar Historical Hydraulic System (Iran):[4] Dating back to the 3rd century BCE "it is as rich in its diversity of civil engineering structures and its constructions as in the diversity of its uses (urban water supply, mills, irrigation, river transport, and defensive system)". Its close relationship to water resources, but also to soil and land-use management, is obvious;

2 For a detailed description see UNESCO, 'West Lake Cultural Landscape of Hangzhou' whc.unesco.org/en/list/1334 last accessed 2 December 2013.

3 For a detailed description see UNESCO, 'Angkor' whc.unesco.org/en/list/668 last accessed 2 December 2013.

4 For a detailed description see UNESCO, 'Shushtar Historical Hydraulic System' whc.unesco.org/en/list/1315 last accessed 2 December 2013.

Figure 1: Cultural Heritage sites showing their close relation to their natural environment and environmental resources: West Lake Cultural Landscape of Hangzhou (China); photo credits: shutterstock.

Figure 2: Cultural Heritage sites showing their close relation to their natural environment and environmental resources: Angkor (Cambodia); photo credits: shutterstock.

Figure 3: Cultural Heritage sites showing their close relation to their natural environment and environmental resources: Shushtar Historical Hydraulic System (Iran); photo credits: shutterstock.

Figure 4: Cultural Heritage sites showing their close relation to their natural environment and environmental resources: Historic Centre of Vienna (Austria); photo credits: shutterstock.

- Historic Centre of Vienna (Austria):[5] While the justification for inscription focuses on the architectural qualities, it is clear that the river Danube is an essential element within the city centre.

Looking beyond the official World Heritage list there are more examples where the sustainable use of water and soil resources has inspired people to create remarkable technical constructions and inventions, which typically do have a strong cultural dimension. One such example are so-called Qanats, historic hydraulic structures developed by the Iranian people, dating back to the 1st millennium BC. A Qanat is a series of well-like shafts, connecting gently sloping underground channels transporting water over long distances to human settlements and irrigated land. This technology is widely used in arid and semi-arid areas. A category II UNESCO Centre was established in 2005 and is active in research, training and technology transfer related to Qanat technology and other historic hydraulic structures.[6]

5 For a detailed description see UNESCO, 'Historic Centre of Vienna' whc.unesco.org/en/list/1033 last accessed 2 December 2013.

6 ICQHS, Official Homepage (International Center on Qanats and Historic Hydraulic Structures) www.icqhs.org/English/Default.aspx accessed 2 December 2013.

In conclusion: virtually all Cultural Heritage Sites (besides other technical and cultural monuments worth preserving) similarly to Natural Heritage sites rely on – and are closely related to – water and soil resources of their environment. This means they cannot be managed and maintained without considering their natural environment and its environmental resources, in particular water and soil.

II Impact of Global Change

Typical challenges for the ecosystems of Natural Heritage sites mostly refer to illegal human activities such as hunting, fishing, clearance of woods or any type of pollution. Climate change poses another challenge by changing temperature and precipitation patterns, increasing frequency and intensity of catastrophic events affecting water and soil resources and associated phenomena such as sea level rise, desertification etc. Climate change has direct and indirect effects on the flora and fauna of Natural Heritage sites and may cause species' extinctions, shifts in species' geographic distributions, increasing frequency of neobiota and alterations in ecosystem structure and function due to changes in species' interactions.[7]

In addition, however, climate change may increase the pressure on Natural Heritage sites by increasing human impact, since the people in the region may face a shortage of natural resources due to climate-driven changes in water and food security, desertification, caused both by decreased rainfall and intensified land use, increased frequency and intensity of floods and droughts etc. Increasing population densities and urbanization put additional pressure on environmental resources and are intricately linked to climate change.

Overall, due to this close inter-linkage and concomitance of factors, one has to consider impact of global change, including climate, but also population growth, urbanization and demographic changes when looking at the challenges for integrated resources management.[8]

---

7 Céline Bellard et al, 'Impacts of Climate Change on the Future of Biodiversity' (2012) 15 Ecology Letters 365.

8 UNEP, 'Status Report on the Application of Integrated Approaches to Water Resources Management' (UNEP 2012).

Figure 5: Impact of Global Change (various aspects) on World Heritage Sites (natural and cultural) and on Environmental Resources. Climate change affects World Heritage both directly and indirectly via environmental resources, which are also affected by other aspects of global change. An integrated management approach for World Heritage therefore needs to include measures targeting them directly, but also the management of environmental resources. It also includes capacity development and governance issues addressing the aspects of global change.

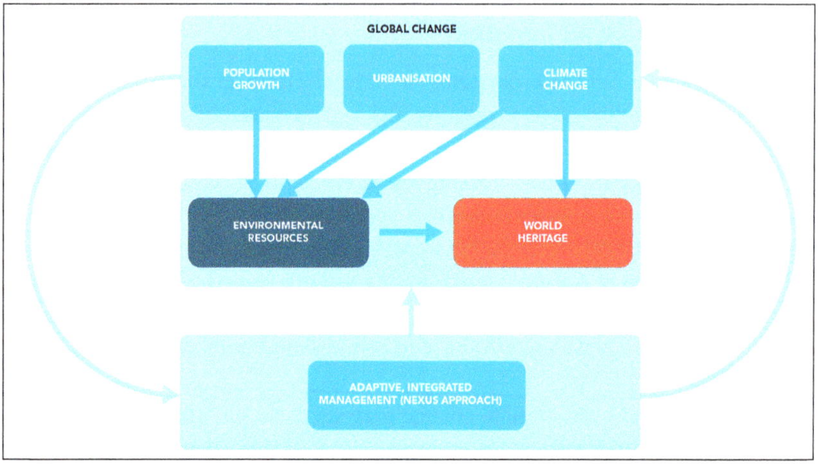

For Cultural Heritage sites, similar to natural ones, there will also be a direct impact of climate, via precipitation and temperature patterns, related both to extreme events[9] and long-term changes in temperature and precipitation patterns.[10] Technical measures to preserve and protect buildings, monuments etc may be applied to mitigate these direct effects. As laid out in the previous section, Cultural Heritage sites in the majority of cases cannot be considered, maintained and managed without their natural environment, in particular with respect to water and land-use. Therefore, impacts of climate change on World Heritage sites (both cultural and natural, see above) will to a large extent be transferred via environmental resources (see upper part of *Figure 2*). These are, besides climatic factors, strongly influenced by other aspects of global change, in particular population growth and urbanization.

---

9 See the contribution of Michael Turner & Rachel Singer on *Urban Resilience in Climate Change.*

10 See the contribution of Roger-Alexandre Lefèvre on *The Impact of Climate Change on Slow Degradation of Monuments in Contrast to Extreme Events.*

III Managing Environmental Resources

Given the close relationship of World Heritage sites to their environment, managing environmental resources, in particular water and soil, will be an important issue for maintenance, development and management of World Heritage. Looking at common approaches to manage water and soil resources, a sectorial view on their management was and still is usually practised, typically also reflected in the respective bureaucracy and organisational structure of ministries (governance) as well as in education.

1. Water Management as an Example

In the case of water management, responsibilities are often distributed between various sectors: energy (hydropower, cooling water), agriculture (irrigation), health (drinking water supply), environment, fisheries, public works/transport (water ways and other infrastructure) etc. The fragmented and sectorial view often resulted in conflicting management strategies and overall poor management results. As a response to this, the concept of Integrated Water Resources Management (IWRM) was developed and has become widely accepted in recent years. It aims at considering water across all compartments and phases, taking into account the various uses and users and also the institutional framework, governance and capacity development.

IWRM was in particular promoted and emerged from the 1992 Earth Summit in Rio de Janeiro, which addressed development within an environmental framework and thus emphasized the global perspective of IWRM. Many research projects and case studies have since been conducted around the world and have provided some general lessons, but many challenges remain to the application of IWRM in practice.[11] Not surprisingly, the implementation of the approach into water governance lags behind both in many developed countries, as well as in developing countries, although it has at least been initiated.[12] By nature, IWRM involves and addresses to a certain extent waste (water) management in addition to soil and land use management as far as land use (e.g. in a river basin) affects water quantity and quality. However, IWRM looks at soil and waste only from a water perspective.

---

11 Roberto Lenton and Mike Muller (eds), *Integrated Water Resources Management in Practice: Better Water Management for Development* (Earthscan 2009).

12 UNEP (n 8).

Conversely, soil and waste *integrated* management approaches have also been propagated. While they provide a wider view on the issues than a purely technical/engineering perspective (which has been common in earlier decades), they fail to provide a holistic approach. Neither IWRM nor the concepts of integrated soil (fertility) management or integrated waste management capture and consider the full range of inter-connectivity and the feed-back loops between these resources, which would be needed to achieve sustainable management. Therefore, it is nowadays increasingly recognized that they should be developed further and merged to an integrated resources management approach, an approach explicitly considering the nexus of these resources.

2. The nexus approach to management of environmental resources

With specific reference to interlinked environmental resources, the term 'nexus' has already been used in the 1980s within the framework of the Food Energy Nexus Programme of UNU.[13] However, it was only in – and after – the Bonn 2011 Conference on "The Water, Energy and Food Security Nexus – Solutions for the Green Economy" that the concept and the term nexus itself gained wider recognition.[14] This approach emphasizes the interdependence of water, energy and food security, taking synergies and trade-offs in the management of these nexus elements into account. Various case studies provide evidence that a nexus approach may facilitate integrated management and governance strategies, overall promoting sustainability and the transition to a Green Economy.

From a resources management perspective on the nexus of water, energy and food security, a nexus approach needs to be developed and implemented by integrating water resources management and soil and land-use management. Also to be considered and included in the nexus approach is waste management, especially concerning organic matter and the recycling of nutrients.[15] The nexus of water, soil and waste results from various material flows and transitions, e.g. the soil transferring run-off and percolation into the *blue* water (surface and groundwater), and the latter by soil into the *green* water that plants use for their growth. The soil also contributes to the transformation of waste from crops, animals and trees into humus and plant nutrients and vice versa. Human use of

13 Ignacy Sachs and Dana Silk, *Food and Energy: Strategies for Sustainable Development* (United Nations University Press 1990).

14 Holger Hoff, 'Understanding the NEXUS' Background Paper for the Bonn 2011 Conference 'The Water, Energy and Food Security Nexus' (Stockholm Environment Institute 2011).

15 Rattan Lal, *The Nexus of Soil, Water and Waste* (UNU-FLORES 2013).

water creates *grey* and *black* waters, which can be used through purification as a source of water and plant nutrients. Also, the application of sludge on soil is an important source of nutrients and organic matter. The contaminated water, grey and black, must be converted into blue and green by denaturing and filtration through soil. The goal is reuse and recycling of the waste following appropriate purification treatment. Thus, the nexus of water, soil and waste is essential for their sustainable management, for increasing water efficiency and soil productivity for food production and for adapting to climate change (see below).

The sustainable management strategies to be developed, advancing a nexus approach, have to be based on consistent and comprehensive systems and flux analysis approaches. Flux analysis, the quantification of material flows and the consistent tracing and follow up of the resource under question, throughout its migration (passage, flow, transport, transfer), through subsequent compartments and phases is essential for closing cycles and a prerequisite for sustainable management. In case of the resource water, this approach would imply to close both the so-called small and large water cycles. The small one describes the sequence *nature* withdrawal, channel transport of water, treatment plant, distribution system, consumption/use, collection and transport of sewage and waste water, treatment, recycling and/or return to *nature*, and thus it is linked to the large water cycle (passage of water in the *natural* hydrosphere: atmosphere, biosphere, lithosphere, oceans and so forth). Besides closing cycles, the nexus approach requires linking cycles, in particular considering the linkage between water, soil and waste and the associated materials as briefly outlined above.

Obvious links arise from the fact that water is not only a resource, but also a transport medium and dissolver of minerals such as nutrients. Taking the case of phosphors, a limited resource and an essential element in all organisms and an important nutrient in agriculture, it clearly represents a broken biogeochemical cycle.[16] While peak phosphorus is lying ahead, too much of this essential element is *lost* via erosion and run-off to rivers, lakes and ultimately the sea, causing eutrophication problems and making recycling unfeasible, at least with current technologies. Systems and flux analysis approaches, using modelling tools are required to quantify material flows and losses and to identify the gaps in the cycle. Respective models may also be used for scenario analysis to study the impacts of global change, considering for example climate change as well as urbanization, population growth and demographic changes.

16 James Elser and Elena Bennett, 'Phosphorus Cycle: A Broken Biogeochemical Cycle' (2011) 478 Nature 29.

3. Adaptive Management

Environmental resources management always needed and needs to cope with uncertainties and changing boundary conditions, but the challenges are becoming more pressing in times of concurrent global trends in terms of climate, population growth, urbanization etc and correspondingly accelerating material flows. As an example, not only does climate change accelerate the *large* hydrological cycle, but also globalized markets accelerate the flow of virtual water around the world. This issue cannot be neglected in the context of water management and offers opportunities for adaption via economic incentives and regulations. Another driver of the water cycle is land-use change, e.g. deforestation affecting evapotranspiration patterns. Conversely, land-use management can be used as a tool for water management. Within a nexus approach, this could be combined with other aspects of *climate-smart* agriculture, which has been promoted in recent years.[17] Changing land-use and agricultural practices may also enhance carbon sequestration as one aspect of climate change mitigation. Adaptation to climate change through a nexus approach may also be applied in urban areas, as exemplified in the blue green dream initiative.[18]

Coming back to World Heritage and their adaptive management in the context of their surrounding environment and its resources, a nexus approach to water, soil and waste management offers more opportunities for adaptive management than any sector-oriented management approach. This will, besides management measures targeted directly to World Heritage sites, address the indirect effect of global change via environmental resources (lower part of *Figure 2*).

A nexus perspective on environmental resources management not only needs to have a holistic view on the material flows and cycling of water, soil and waste, but also needs to be policy oriented. This was one of the main items worked out in the white book on Advancing a Nexus Approach to the Sustainable Management of Water, Soil and Waste.[19] How to implement a nexus approach to environmental resources management, which institutional frameworks will be best suited and how they can be developed and improved are among the main questions to be addressed. Closely related to this issue is the need for capacity

---

17 FAO, 'Coping with a Changing Climate: Considerations for Adaptation and Mitigation in Agriculture' (Rome 2009).

18 Blue Green Dream Project, Official Website bgd.org.uk/accessed 9 December 2013.

19 UNU-FLORES, 'White Book – Advancing a Nexus Approach to the Sustainable Management of Water, Soil and Waste' (UNU-FLORES 2014).

development, taking a multi-level approach.[20] To be able to implement a nexus approach, well-trained stakeholders at all levels and all sectors (academia, practitioners, decision makers), which understand the concept, are required. This can be addressed by capacity development at the individual level, requiring target and region-specific study programmes as well as training programmes and courses addressing the nexus of water soil and waste. Successful implementation of management strategies requires also, however, an enabling environment, which is an issue in institutional capacity development.

4. UNU-FLORES: A Response to the Challenges

Since the Bonn 2011 Nexus Conference and partly in parallel to it, various initiatives and projects have been started to carry and develop the nexus concept further. One such initiative is the establishment of the United Nations University Institute for Integrated Management of Material Fluxes and of Resources (UNU-FLORES) in Dresden, Germany. It was initiated and proposed jointly by UNU and its partner university TU Dresden. UNU-FLORES was officially inaugurated in December 2012. Its mission is directly related to the nexus concept, namely "to contribute, through research, teaching, advanced training, capacity development and knowledge dissemination to the resolution of pressing challenges in the area of integrated management of environmental resources: soil, water and waste that are of concern to the United Nations and its member states particularly in developing and emerging economies".[21] UNU-FLORES aims at acting at the forefront of initiatives promoting a nexus approach by serving within the UN system as a think tank that is internationally recognized as a major hub and intellectual focal point, promoting integrated management of environmental resources.

UNU-FLORES is unique in the sense that its academic functional structure is consistent with the water-soil-waste nexus (see *Figure 3*), which is, as outlined above, closely related to the water, energy and food security nexus promoted by the Bonn 2011 conference. The organization of UNU-FLORES into

---

20 Marco Leidel, Steffen Niemann and Nina Hagemann, 'Capacity Development as a Key Factor for Integrated Water Resources Management (IWRM): Improving Water Management in the Western Bug River Basin, Ukraine' (2012) 65 Environmental Earth Sciences 1415.

21 UNU, 'Statute of the UNU Institute for Integrated Management of Material Fluxes and of Resources (UNU-FLORES)' (United Nations University 2010) unu.edu accessed 19 March 2014.

five academic units – three core scientific units dealing with the interconnected resources (Water Resources Management (WRM), Waste Management (WM) and Soil and Land use Management (SLM) supported by two cross cutting units (Systems and Flux Analysis considering global change assessment (SFA) and Capacity Development and Governance (CDG) – supports the think tank function of UNU.

Figure 6: Academic functional structure of UNU-FLORES, reflecting the nexus approach to the management of water, soil and waste.

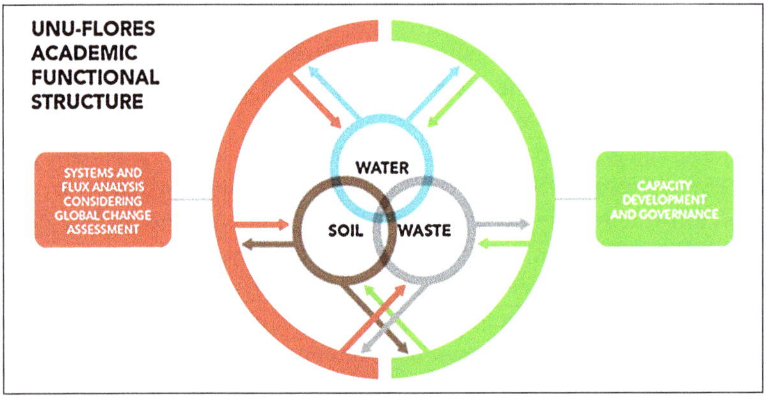

The work program of core scientific units to a large extent will be based on close cooperation with SFA to better understand the interactions between atmosphere, biosphere, hydrosphere, lithosphere and pedosphere. It is envisaged that the capacity development and governance unit will create demand for knowledge products (such as online courses), identify opportunities to field test new methodologies, facilitate cross-fertilization of ideas across regions based on institutional good practice and seek to create partnerships for education, research and training that support a think tank function.

UNU-FLORES will pursue its activities in close interaction with a network of partners. Besides collaborating with other UNU institutes working in related areas, it has established close relationships with UN organizations working in the respective areas, such as FAO, UNEP, UN-HABITAT and UNESCO-IHE. A major partner of UNU-FLORES is Dresden University of Technology (TU Dresden), in particular the faculty of environmental sciences, which has ample experience in closing cycles related to the management of water, soil and waste. Major joint activities related to advancing a nexus approach to the sustainable management of water, soil and waste include:

- preparations for a joint PhD programme; its start is planned for October 2014;
- the establishment of a regular international Nexus conference in Dresden; a kick-off workshop took place in November 2013, the major conceptual background had been drafted in a white book[22] which was finalized in early 2014;
- various initiatives for research projects dealing with various aspects of the water-soil-waste nexus.

Another partnership is currently being established in Maputo, Mozambique. This initiative started concomitantly to the establishment of UNU-FLORES in Dresden and aims at creating a regional hub for environmental resources management in Southern Africa. Based in Maputo, closely cooperating with the Eduardo Mondlane University and the Ministry of Science and Technology of Mozambique, but also with a network for research and education spread all over Africa[23], an operating unit of UNU-FLORES will take up activities in 2014.

Building on these and additional partnerships with international research organizations and universities as well as on direct contacts to governmental bodies in member states, UNU-FLORES is well positioned to fulfil its envisaged role as a think tank for integrated management of environmental resources – soil, water and waste – and to consolidate the scientific foundation of the nexus approach by filling critical knowledge gaps relating the nexus approach and its implementation. With its focus on policy-relevant research, considering capacity development and governance, UNU-FLORES will be able to provide the adaptive management strategies as depicted in *Figure 2* as one important component of an integrated management approach to World Heritage sites. Given its mission to act as a think tank for the UN system, UNU-FLORES is keen to cooperate with the UNESCO World Heritage Centre on implementing and adapting such integrated management plans, including short, medium and long-term actions to protect and maintain World Heritage, as outlined in the Operational Guidelines for the Implementation of the World Heritage Convention.[24] Such management systems of World Heritage, considering buffer zones, include a monitoring plan,

22 UNU-FLORES (n 19).

23 Stephan Hülsmann and Reza Ardakanian, 'Proceedings of the Regional Workshop on Establishment of a Network for Partnership of UNU-FLORES Based in Maputo, Mozambique' (UNU-FLORES 2013).

24 UNESCO, 'Operational Guidelines for the Implementation of the World Heritage Convention' (June 2013) WHC. 13/01 whc.unesco.org/en/guidelines accessed 27 January 2014.

capacity building activities and may allow the use of World Heritage properties in a sustainable way. This might best be achieved by developing a nexus approach to the management of water, soil and waste for the respective World Heritage sites and implementing it in the framework of the management system for these properties. The respective state parties, which are requested to submit the periodic reports on the status of World Heritage, should, via intergovernmental mechanisms, be involved in environmental resources management.

IV Summary and Conclusion: Integrated Management of Environmental Resources as Means to Manage and Maintain World Heritage

Mitigating the direct and indirect impacts of climate change on World Heritage (both natural and cultural) requires an integrated approach. Protection and management of World Heritage must be considered in the context of managing environmental resources in the region in a sustainable way. The most appropriate approach to achieve this should be a nexus approach considering the close interrelations of environmental resources such as water, soil and waste. In addition, especially in developing countries, these management strategies need to be embedded in a suite of measures and activities in capacity development addressing all relevant stakeholders, thus considering both individual and institutional capacity development. UNU-FLORES considers itself as a direct response of the UN system to address these issues and to advance a nexus approach to the sustainable management of water, soil and waste, acting as a think tank for the UN system and member states through policy-relevant research, education and capacity development. Implementing a nexus approach in the respective region or city will be one means and an important component to manage and maintain World Heritage, be it natural or cultural.

Bibliography

Bellard, C, Bertelsmeier C, Leadley P, Thuiller W and F Courchamp, 'Impacts of Climate Change on the Future of Biodiversity' (2012) 15 Ecology Letters 365

Blue Green Dream Project, Official Website bgd.org.uk/accessed 9 December 2013

Elser, J and E Bennett, 'Phosphorus Cycle: A Broken Biogeochemical Cycle' (2011) 478 Nature 29

FAO, 'Coping with a Changing Climate: Considerations for Adaptation and Mitigation in Agriculture' (Rome 2009) www.fao.org/docrep/012/i1315e/i1315e00.htm accessed 5 December 2013

Hoff, H, 'Understanding the NEXUS, Background Paper for the Bonn2011 Conference: The Water, Energy and Food Security Nexus' (Stockholm Environment Institute 2011) www.water-energy-food.org/en/whats_the_nexus/background.html accessed 5 December 2013

Hülsmann, S and R Ardakanian, 'Proceedings of the Regional Workshop on Establishment of a Network for Partnership of UNU-FLORES Based in Maputo, Mozambique' (UNU-FLORES 2013)

ICHQS, Official Website www.icqhs.org/English/Default.aspx accessed 2 December 2013

Lal, R, 'The Nexus of Soil, Water and Waste' (UNU-FLORES 2013) flores.unu.edu/publications accessed 6 December 2013

Lefèvre, R-A, 'The Impact of Climate Change on Slow Degradation of Monuments in Contrast to Extreme Events', *Climate Change as a Threat to Peace: Impacts on Cultural Heritage and Cultural Diversity* (2014)

Leidel, M, Niemann and N Hagemann, 'Capacity Development as a Key Factor for Integrated Water Resources Management (IWRM): Improving Water Management in the Western Bug River Basin, Ukraine' (2012) 65 Environmental Earth Sciences 1415

Lenton, R and M Muller (eds), *Integrated Water Resources Management in Practice: Better Water Management for Development* (Earthscan 2009)

Sachs, I and D Silk, *Food and Energy: Strategies for Sustainable Development* (United Nations University Press1990)

Turner, M, 'Urban Resilience for Climate Change', *Climate Change as a Threat to Peace: Impacts on Cultural Heritage and Cultural Diversity* (2014)

UNEP, 'Status Report on the Application of Integrated Approaches to Water Resources Management' (UNEP 2012) www.un.org/waterforlifedecade/pdf/un_water_status_report_ 2012.pdf accessed 5 December 2013

UNESCO, 'Operational Guidelines for the Implementation of the World Heritage Convention' (June 2013) WHC. 13/01 whc.unesco.org/en/guidelines accessed 27 January 2014

UNU, 'Statute of the UNU Institute for Integrated Management of Material Fluxes and of Resources (UNU-FLORES)' (United Nations University 2010) unu.edu accessed 5 December 2013

UNU-FLORES, 'White Book – Advancing a Nexus Approach to the Sustainable Management of Water, Soil and Waste' (UNU-FLORES 2014) flores.unu.edu/white-book/

Elizabeth Longworth*

The Culture of Prevention: Heritage and Resilience

Abstract Cultural heritage is an integral factor of sustainable development, climate change adaptation, disaster risk reduction and community resilience. Today, both tangible and intangible heritage are at risk due to development and socio-economic transformations, natural hazards, disasters, climate change, urbanization, unsustainable tourism, conflicts and political tensions. Laudable efforts are being made at the global, regional, national and local levels to protect cultural patrimony from these threats. However, more focused and strategic action must be taken by all concerned stakeholders; in a spirit of co-operation, to ensure that heritage is protected and that its role in reducing the risks of disaster (including climate risk) and building resilience is recognized and promoted.

I Heritage at Risk

Cultural patrimony manifests in two ways: there is tangible heritage, such as monuments, museums, historic cities, archaeological sites and cultural landscapes; and there is intangible heritage or cultural expressions, such as traditional knowledge, practices, skills and crafts.

Today, both tangible and intangible heritage are at risk due to a number of factors. Natural hazards and disasters (particularly, earthquakes, landslides and floods), climate change, urbanization, unsustainable tourism, conflicts and

* The former director of the United Nations Office for Disaster Risk Reduction (UNISDR). The presentation was made during the UNESCO Conference "Climate Change as a Threat to Peace: Impacts on Cultural Heritage and Cultural Diversity", 31 May 2013, Dresden, Germany, draws largely from "Heritage and Resilience: Issues and Opportunities for Reducing Disaster Risks", a paper produced for the UNISDR – convoked Fourth Session of the Global Platform for Disaster Risk Reduction (19–23 May 2013, Geneva, Switzerland), and the outcomes of the first round of consultations related to the successor arrangement to the Hyogo Framework for Action: Building the Resilience of Nations and Communities to Disasters (2005–2015), that were organized or participated in by the United Nations Office for Disaster Reduction (UNISDR).

political unrest have negatively impacted and contributed to losses of cultural patrimony. In recent years, more heritage has been lost due to natural hazards and disaster events than ever accounted for in the past.

II Building the Resilience of Nations

Cultural heritage plays an important role in supporting sustainable development, disaster risk reduction and building community resilience,[1] by:

- Providing a source of meaning and identity;
- Promoting a sense of belonging for individuals and communities;
- Serving as a means to organize communities and relationships;
- Promoting values and social cohesion through the cultivation of mutual respect, a strong sense of collective purpose, and responsibility to maintain a common good;
- Contributing to spiritual and psychological well-being;
- Serving as a repository and rich source of traditional knowledge on disaster prevention, mitigation and recovery, as well as climate change adaptation;
- Rebuilding a sense of community after disasters; and
- Providing socio-economic benefits to the community, e.g. through tourism.

In recognition of the importance of Cultural Heritage in disaster risk reduction, the international agreement, entitled the *Hyogo Framework of Action for Disaster Risk Reduction: Building the Resilience of Nations and Communities to Disasters (2005–2015) (HFA)*, identified the use of knowledge (particularly, traditional and indigenous knowledge), innovation and education to build a culture of safety and resilience as one of its priority actions. So significant is the role of heritage in community resilience and sustainable development, that the Rio+20 Conference (United Nations Conference on Sustainable Development, 2012) called for its conservation.

1 Resilience is the ability of a system to reduce, prevent, anticipate, absorb and adapt, or recover from the effects of a hazardous event in a timely and efficient manner, including through ensuring the preservation, restoration, or improvement of its essential basic structures and functions, see the United Nations Plan of Action on Disaster Risk Reduction for Resilience, 2013.

Figure 1: Hyogo Framework for Action on Disaster Risk Reduction: Priorities for Action 2005–2015.

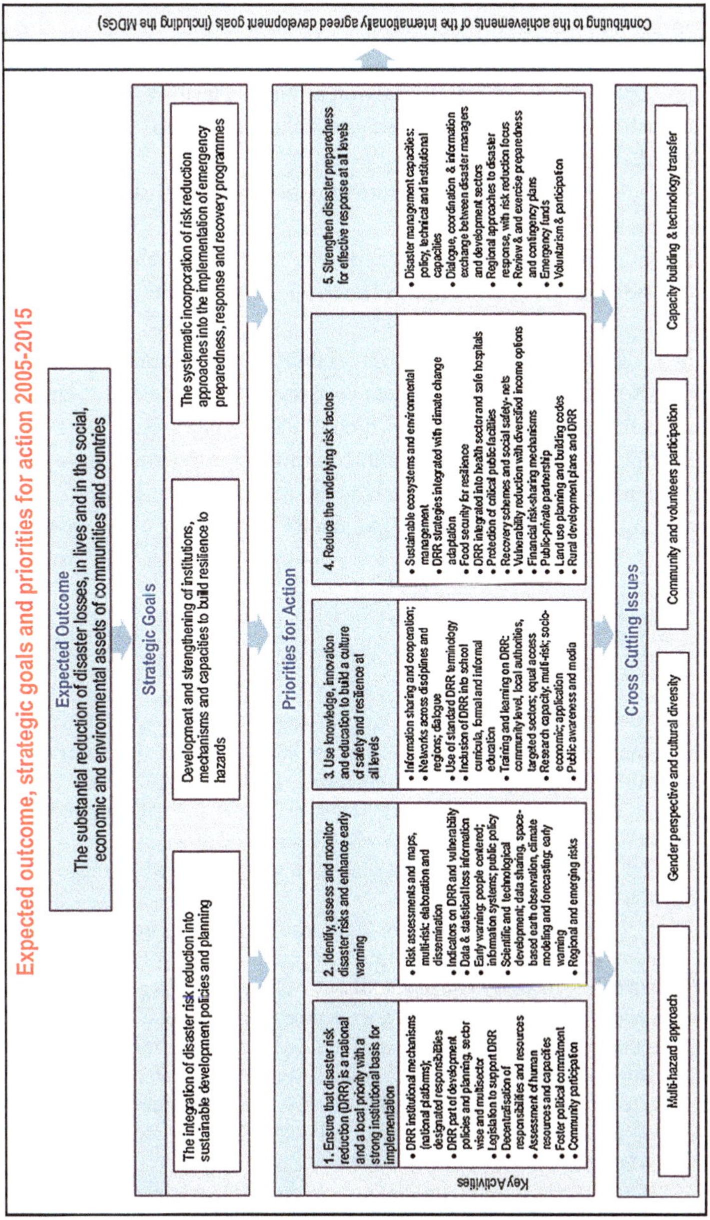

III Cooperation at Work

Laudable efforts have been and are still being made at the global, regional, national and local levels to protect cultural patrimony from threats. At the global level, international conventions (e.g. the World Heritage Convention, 1972[2]), strategies (e.g. Strategy for Risk Reduction at World Heritage, 2007[3]) and frameworks for action (e.g. HFA) have been established; technical resources have also been published for stakeholders.

At the regional level, Regional Platforms for Disaster Risk Reduction[4] have been launched.

At the national level, disaster risk reduction plans and policies have been established and National Platforms for Disaster Risk Reduction[5] have also been launched.

At the local level, networks and partnerships between cities (including twinning initiatives) have been established; conservation and cooperation plans are being pursued (see *Figures 2 and 3*, which illustrate the cases of Petra and Timbuktu); public-private partnerships and community engagement have been promoted; and advocacy campaigns, such as the *Making Cities Resilient: My City Is Getting Ready* Campaign, have been launched.[6]

---

2 Convention concerning the Protection of the World Cultural and Natural Heritage; adopted 16 November 1972, entered into force 17 December 1975, 1037 UNTS 151.

3 UNESCO, 'Climate Change and World Heritage. Report on Predicting and Managing the Impacts of Climate Change on World Heritage and Strategy to Assist States Parties to Implement Appropriate Management Responses' World Heritage Reports 22 (1 May 2007).

4 Regional platforms are multi-stakeholder forums organized by UNISDR and host governments that reflect the commitment of countries to improve coordination and implementation of disaster risk reduction activities while linking to international and national efforts, see UNISDR website, www.unisdr.org, accessed 20 March 2014.

5 National Platforms are nationally owned and led multi-stakeholder forums or committees working on disaster risk reduction. They reflect the commitment of its government to implement national and local disaster risk reduction activities while linking up to international efforts, see UNISDR website, www.unisdr.org, accessed 20 March 2014.

6 UNISDR, 'Making Cities Resilient', www.unisdr.org/campaign/resilientcities/, accessed 20 March 2014.

Figure 2: Petra: Conservation and cooperation in action

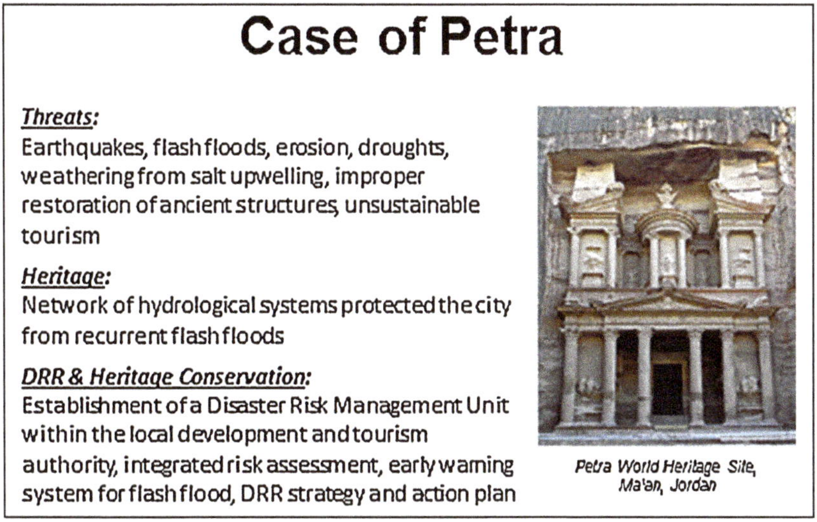

Figure 3: Timbuktu: Conservation and cooperation in action

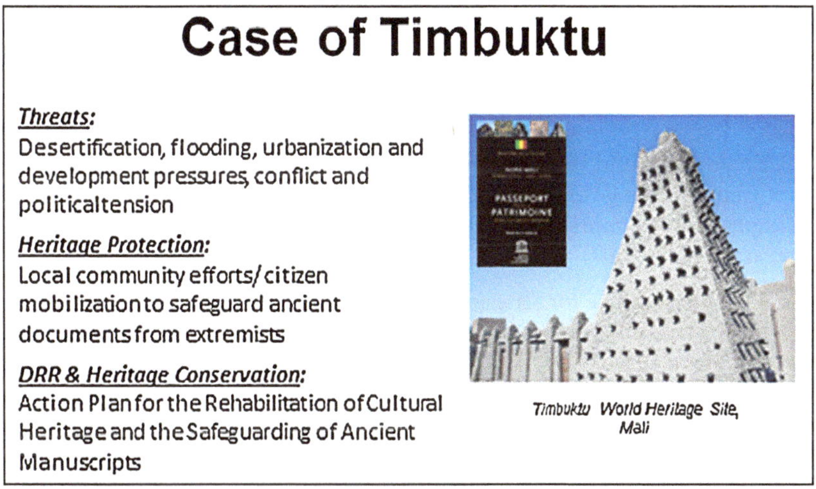

IV Heritage-driven Resilience

All concerned stakeholders, in a spirit of cooperation, must take more focused and strategic action to ensure that heritage is protected and that its role in disaster risk reduction and climate change adaptation is recognized and promoted.

Action must be taken to support efforts towards heritage-driven resilience, in particular through the:

- Integration of heritage in disaster risk reduction plans, policies and strategies at all levels;
- Development of awareness and capacity-building programmes, e.g. by establishing linkages between the Making Cities Resilient Campaign and the World Heritage Cities;
- Harnessing of public-private partnerships; and
- Strengthening of existing partnerships and networks[7] in order to set the agenda and provide normative guidance, ensure coordinated action and link technical expertise with the needs.

Two major opportunities currently exist for developing a global culture of prevention. One is the international debate on the post-2015 international development agenda; the other is the consultation and drafting process for the successor arrangement to the *Hyogo Framework for Action for Disaster Risk Reduction* (culminating in the World Conference on Disaster Risk Reduction in Sendai, Japan, in 2015). Governments and citizens alike should capitalize on these opportunities and make decisions to protect their heritage while ensuring that nations as well as present and future generations are resilient to disasters.

Bibliography

United Nations Educational, Scientific and Cultural Organization, 'Climate Change and World Heritage. Report on Predicting and Managing the Impacts of Climate Change on World Heritage and Strategy to Assist States Parties to Implement Appropriate Management Responses' World Heritage Reports 22 (Paris, France 2007)

United Nations Office for Disaster Risk Reduction, 'Proceedings of the Fourth Global Platform for Disaster Risk Reduction' (19–23 May 2013) (Geneva, Switzerland 2013)

–, 'Synthesis Report on Consultations on the Post-2015 Framework on Disaster Risk Reduction (HFA2)' (Geneva, Switzerland 2007)

7 Partnerships and networks exist between the following: heritage managers; Regional and National Platforms for Disaster Risk Reduction; National Disaster Management Organizations; political leaders; private sector actors (small and medium-sized enterprises); and technical institutes.

–, 'Hyogo Framework for Action: Building the Resilience of Nations and Communities to Disasters 2005–2015: Extract from the Final Report of the World Conference on Disaster Risk Reduction' (Geneva, Switzerland 2007)

–, 'Making Cities Resilient' (Geneva, Switzerland 2012)

–, UNESCO World Heritage Centre, International Council on Monuments and Sites – International Scientific Committee on Risk Preparedness, International Centre for the Study of the Preservation and Restoration of Cultural Property and MARSH International, 'Heritage and Resilience: Issues and Opportunities for Reducing Disaster Risks' (Geneva, Switzerland 2013)

United Nations System Chief Executives Board for Coordination (CEB), 'UN Plan of Action on Disaster Risk Reduction for Resilience' (Geneva, Switzerland 2013)

Lex Lata and *De Lege Ferenda* – Legal Challenges of Cultural Property Protection in the Context of Climate Change

Guido Carducci*

What Consideration is Given to Climate and to *Climate Change* in the UNESCO Cultural Heritage and Property Conventions?

I A Preliminary Clarification: Climate *per se*, the Protection of Cultural Property and Heritage versus Climate Change

Before entering into the rather recent debate on climate change, which is a rapid, and in part unexpected, increase in the worldwide temperature, it is methodologically appropriate to make a few remarks as to climate *per se* and the protection of cultural property and heritage law.

Going beyond any misleading appearance of a monolithic composition of property and heritage, *cultural property* – and even more if it is combined with *Cultural Heritage* – includes a huge variety of types and categories of movables and immovables. Each of them is made of materials that are more or less climate-sensitive and that deserve to be considered individually; conserved and protected in different ways. There is no one-fits-all solution and the Cultural Heritage or property material is, from a climate perspective, as relevant as the heritage or the property itself. For instance, humidity and rain can affect and heavily damage wood (as in structures, doors, frames), textiles (as in carpets) and untreated paper (as in maps and drawings), although it does not affect equally stone in a sculpture. Each category of material that composes cultural property and heritage is to be considered individually. Each has its own degree of sensitiveness to climate and of fragility to climate-related events, such as humidity, rain, draught, etc.

Climate *per se* naturally changes over seasons. However, the so-called global *climate change* is something different. It is a new, much greater in scale and faster in growth, phenomenon than the ordinary change of climate over seasons. Reasons for such global climate change vary and include human activities, such as industrial development and the burning of fossil fuels like coal and oil.

---

* Former Chief, Legal and Treaty Section (International Standards), UNESCO, Law Professor (Maître de Conférences, Univ. Paris Est; Visiting Professor Univ. of Miami and of Rome (Luiss)); Ph.D.s Univ. Paris II and Rome I; Diploma Hague Academy of International Law; Attorney; Chartered Arbitrator; FCIArb (gcarducci@noos.fr).

These change the natural greenhouse emissions which keep heat radiating from Earth from leaving the atmosphere. The result is an increase in temperature with various consequences, from loss of sea ice and accelerated sea level rise to long draughts and stronger typhoons.

II Climate Change and the Legal Protection of Cultural Property and Heritage

1. Law versus Climate Change

Cultural property and heritage are affected by global climate change as is anything else on earth which is composed of the same material with the same degree of fragility for the type of climate change concerned, such as increasing rain falls, rising sea level, etc. The fact that in many countries Cultural Heritage and property is often protected by legislation does not, *per se*, exclude the climate change threats. First, with regard to immovable Cultural Heritage, the law cannot remove it from its *situs* and protect it fully from the impact of nature. Second, with regard to movable cultural property, the law can organize a removal of the property from its *situs* and its gathering in museums, which does protect from most, not all, impacts of nature. However, museums can hardly protect from flooding or other serious effects of climate change. For instance, some museums in the Czech Republic had to face unexpected floods. Again, the specificity and resistance of each material does matter. For example, book collections are naturally more affected by floods than marble columns or statues. Further, they require cleaning and drying in addition to being disinfected if dangerous bacteria were in the flood water. Preparedness to deal with emergency situations is a must for most museums, foundations and collections. Standards have increased over the years and today such measures include floods and any other emergency situation, whether or not they are climate change-related.

But climate change can do more than flooding a territory, including museums and collections. From a legal perspective, it could go as far as cancelling a territory which is one of the constitutive elements of a State. A flat and small island could be submerged in a few years or decades; its territory disappears, becoming an underwater surface. At law, a State generally does not exist without a territory. Also, international law is based on States with their territories, population and governments. Without such elements a State no longer exists at law. If a population leaves, the State disappears. Any cultural property still on the territory would self-evidently no longer be legally protected by national legislation nor by international law as the latter no longer regards the territory as part of a State bound by its rules. The State might appear again; nevertheless, if the sea level would decrease, allowing a territory to be re-established with its population and government.

All forms of international legal protection of cultural property and heritage require a State that is bound by either a UNESCO Convention (or other similar conventions) that it has ratified or its own national legislation or is invited by a UNESCO declaration or recommendation to take action and ensure protection of the heritage on its territory. When one takes a closer look at *protection* this term is open-ended. It hides a variety of forms and effects, from measures of legal protection to those that also include material protection. Unsurprisingly, they entail a variety of thresholds in protection. With regard to those threats that climate change generally generates, the relevant forms of legal protection are limited to those legal measures that also include material protection. Declaring some heritage inalienable, for instance, does not protect it from climate change or from the ordinary and less aggressive and unpredictable annual climate change.

At the very least, climate change makes the protection of cultural property and heritage more needed and more expensive. An increasingly larger number of objects are affected by an increase of temperature and/or of humidity, and their conservation requires more museums to be built and/or, in the existing museums, more suitable equipment in terms of air conditioners, temperature stabilizers and dehumidifiers. Such costs are unproblematic for the national authorities of some countries, while problematic for other countries. Costs increase dramatically and prevention becomes more problematic when the risk is flooding, typhoons, and destruction of buildings where the museum or the collection is located.

2. Intangible Cultural Heritage Convention

Intangible Cultural Heritage is the most precarious form of heritage because it cannot rely *per se* on a tangible body, be it movable or immovable, that naturally lasts over time, at least some time. The UNESCO 2003 Convention for the Safeguarding of the Intangible Cultural Heritage has currently reached the remarkable number of 161 States Parties.[1] The terminology of *intangible* Cultural Heritage is still problematic in the countries where for many centuries the term of the tangible Cultural Heritage – be it movable or immovable – has been applied in a monopolistic manner both *de facto* and *de jure*. The terminology of *intangible* Cultural Heritage is unproblematic and rooted in substance, stemming from an anthropological approach to culture of a community.

---

1 Adopted 17 October 2003, entered into force 20 April 2006, 2368 UNTS 3.

The international community gathered at UNESCO had to face these challenges, and many others, that the intangible dimension of culture raises, especially for the purposes of creating an international normative instrument. At the end of the negotiations, the result achieved reads as follows:

> For the purposes of this Convention,
> 1. the "intangible cultural heritage" means the practices, representations, expressions, knowledge, skills – as well as the instruments, objects, artefacts and cultural spaces associated therewith – that communities, groups and, in some cases, individuals recognize as part of their cultural heritage. This intangible cultural heritage, transmitted from generation to generation, is constantly recreated by communities and groups in response to their environment, their interaction with nature and their history, and provides them with a sense of identity and continuity, thus promoting respect for cultural diversity and human creativity. For the purposes of this Convention, consideration will be given solely to such intangible cultural heritage as is compatible with existing international human rights instruments, as well as with the requirements of mutual respect among communities, groups and individuals, and of sustainable development.
> 2. The "intangible cultural heritage", as defined in paragraph 1 above, is manifested *inter alia* in the following domains:
> (a) oral traditions and expressions, including language as a vehicle of the intangible cultural heritage;
> (b) performing arts;
> (c) social practices, rituals and festive events;
> (d) knowledge and practices concerning nature and the universe;
> (e) traditional craftsmanship.[2]

These practices, representations, expressions, knowledge and skills that communities and groups and at times individuals, recognize as part of their intangible Cultural Heritage can be easily put in danger by climate change. For instance, higher temperatures and longer periods of drought can rarefy and even extinguish some types of plants that were used in traditional knowledge and know how, either as medicine or as raw material.

For instance, in its Decision 7.COM 8.3 the Intergovernmental Committee for the Safeguarding of the Intangible Cultural Heritage[3], the Committee first took note of the following, an information which provides the reader some context:

> "Noken is a knotted net or woven bag handmade from wood fibre or leaves by communities in Papua and West Papua Provinces of Indonesia. Men and women use it for carrying plantation produce, catch from the sea or lake, firewood, babies or small

---

2 Article 2.

3 Established under Article 5 of the Convention.

> animals as well as for shopping and for storing things in the home. Noken may also be worn, often for traditional festivities, or given as peace offerings. The method of making Noken varies between communities, but in general, branches, stems or bark of certain small trees or shrubs are cut, heated over a fire and soaked in water. The remaining wood fibre is dried then spun to make a strong thread or string, which is sometimes coloured using natural dyes. This string is knotted by hand to make net bags of various patterns and sizes. The process requires great manual skill, care and artistic sense, and takes several months to master. The number of people making and using Noken is diminishing, however. Factors threatening its survival include lack of awareness, weakening of traditional transmission, decreasing numbers of craftspeople, competition from factory-made bags, problems in easily and quickly obtaining traditional raw materials, and shifts in the cultural values of Noken".[4]

Among these factors threatening survival, "competition from factory-made bags, problems in easily and quickly obtaining traditional raw materials" are likely to be the result of global warming, probably jointly with other causes. The making of Noken often relies on branches, stems or barks of some kind of small trees that are heated over a fire and then soaked in water. Such small trees grow under certain conditions.

The Committee has decided to inscribe Noken, a multifunctional knotted or woven bag which is a handcraft of the people of Papua, on the List of Intangible Cultural Heritage in Need of Urgent Safeguarding,[5] and one of the considerations is:

> "The traditional know-how related to the Noken is in need of urgent safeguarding because of risks of a gap in transmission to younger generations, competition from

4 UNESCO, Intergovernmental Committee for the Safeguarding of the Intangible Cultural Heritage Decision 7.COM 8.3 (3–7 December 2012) ITH/12/7.COM/Decisions, Paragraph 1.

5 Article 17 – List of Intangible Cultural Heritage in Need of Urgent Safeguarding:
1. With a view to taking appropriate safeguarding measures, the Committee shall establish, keep up to date and publish a List of Intangible Cultural Heritage in Need of Urgent Safeguarding, and shall inscribe such heritage on the List at the request of the State Party concerned.
2. The Committee shall draw up and submit to the General Assembly for approval the criteria for the establishment, updating and publication of this List.
3. In cases of extreme urgency – the objective criteria of which shall be approved by the General Assembly upon the proposal of the Committee – the Committee may inscribe an item of the heritage concerned on the List mentioned in paragraph 1, in consultation with the State Party concerned.

> modern and imported products, and the scarcity of traditional materials that are being replaced by synthetic materials".[6]

Besides this Committee's decision, in order to facilitate the safeguarding of the intangible Cultural Heritage, the Convention has established an Intangible Cultural Heritage Fund[7] and a system of international cooperation and assistance.[8]

The forms of such assistance may include: (a) studies concerning various aspects of safeguarding; (b) the provision of experts and practitioners; (c) the training of all necessary staff; (d) the elaboration of standard-setting and other measures; (e) the creation and operation of infrastructures; (f) the supply of equipment and know-how; (g) other forms of financial and technical assistance, including, where appropriate, the granting of low-interest loans and donations.[9]

While such forms of assistance can satisfy a variety of scenarios and different kinds of threats to the intangible Cultural Heritage, several of these forms can effectively contribute to reducing the threats that climate change generates for the safeguarding of the intangible Cultural Heritage.

a) Conventions on Tangible Heritage

Tangible Cultural Heritage or property is the traditional subject matter of UNESCO standard-setting instruments. Among other instruments, these include:

- Convention on the Protection of the Underwater Cultural Heritage (Paris, 2 November 2001)[10]
- Convention concerning the Protection of the World Cultural and Natural Heritage (Paris, 16 November 1972)[11]
- Convention on the Means of Prohibiting and Preventing the Illicit Import, Export and Transfer of Ownership of Cultural Property (Paris, 14 November 1970)[12]

---

6 UNESCO, Intergovernmental Committee for the Safeguarding of the Intangible Cultural Heritage Decision 7.COM 8.3 (3–7 December 2012) ITH/12/7.COM/Decisions, Paragraph 2 U.2.

7 Article 6.

8 Article 5.

9 Article 21.

10 Entered into force 2 January 2009, (2002) 41 ILM 37.

11 Entered into force 17 December 1975, 1037 UNTS 151.

12 Entered in force 24 April 1972, 823 UNTS 231.

- Convention for the Protection of Cultural Property in the Event of Armed Conflict with Regulations for the Execution of the Convention (The Hague, 14 May 1954)[13] and its First Protocol (The Hague, 14 May 1954) and Second Protocol (The Hague, 26 March 1999)[14]

The size of this article clearly does not allow a detailed analysis of these conventions. Generally, climate *per se* is not a variable in the protection of heritage that is expressly considered and taken into account by these conventions.

b) World Cultural and Natural Heritage

An important exception is provided by the famous Convention concerning the Protection of the World Cultural and Natural Heritage. With 190 States Parties this Convention has the privilege to stand out as a nearly universally applicable convention. Under this Convention, each State Party submits to the World Heritage Committee an inventory of property forming part of the Cultural and Natural Heritage of their State which has outstanding universal value.

In addition, the Committee has the duty to establish, keep up to date and publish, whenever circumstances shall so require, a List of World Heritage in Danger, *id est* a list of the property appearing in the World Heritage List for the conservation of which major operations are necessary and for which assistance has been requested by a State Party. This List of World Heritage in Danger may include only such property forming part of the Cultural and Natural Heritage as is threatened by serious and specific dangers which include, *inter alia*, "threat of disappearance caused by accelerated deterioration (…), destruction caused by changes in the use or ownership of the land; major alterations due to unknown causes; abandonment for any reason whatsoever; (…) calamities and cataclysms; serious fires, earthquakes, landslides; volcanic eruptions; changes in water level, floods, and tidal waves".[15]

It is rather remarkable that this Convention, adopted in 1972 – at a time where climate change was generally not considered *per se*, but simply as part of the broader and general climate annual change – expressly includes weather-related phenomena that would probably be associated with climate change today. However, these causes were codified in 1972. Above all, these provisions do not necessarily provide evidence of what one defines as climate change today. In fact, it has rather been intended as a matter of principles. It has become clear that

13 Enterd into force 7 August 1956, 249 UNTS 240.

14 Entered into force 9 March 2004, (1999) 38 ILM 769.

15 Article 11 (4).

States most affected by one or more causes of climate change and comprising of world Cultural and Natural Heritage sites on their territories are most likely to endorse the insertion of those principles by the Committee on the List of World Heritage in Danger.

This article deals only with climate and climate change in the protection of Cultural Heritage and property, which is a dimension that is generally not considered *per se*. It is generally dealt with as a limited part, not particularly thoroughly examined, of the much broader debate on Managing Disaster Risks, primarily for the World Heritage sites.[16]

c) Underwater Cultural Heritage

Seas are directly affected by climate change in various ways. For instance, their bio system varies naturally with an increase of temperature. Their level rises as a result of greater ice melt. Stronger winds and typhoons make exceptional waves and tsunamis more frequent.

However, the situation on the seabed is, in part, affected differently by climate change than the situation at sea level is. This remark is not irrelevant when one focuses on the Convention on the Protection of the Underwater Cultural Heritage (Paris, 2 November 2001). This Convention is focused on underwater Cultural Heritage, not generally on the law of the sea. What such heritage includes is made clear by the Convention:

> (a) "Underwater cultural heritage" means all traces of human existence having a cultural, historical or archaeological character which have been partially or totally under water, periodically or continuously, for at least 100 years such as:
> (i) sites, structures, buildings, artefacts and human remains, together with their archaeological and natural context;
> (ii) vessels, aircraft, other vehicles or any part thereof, their cargo or other contents, together with their archaeological and natural context; and
> (iii) objects of prehistoric character.
> (b) Pipelines and cables placed on the seabed shall not be considered as underwater cultural heritage.
> (c) Installations other than pipelines and cables, placed on the seabed and still in use, shall not be considered as underwater cultural heritage.[17]

---

16 For instance, see UNESCO, 'Managing Disaster Risks for World Heritage' (2010) World Heritage Resource Manual whc.unesco.org/document/104522 accessed 23 January 2014.

17 Article 1.

Importantly, any thought that whatever is on the seabed is under threat and needs to be removed is defeated by the clear principle that inspires the whole Convention: "(t)he preservation *in situ* of underwater Cultural Heritage shall be considered as the first option before allowing or engaging in any activities directed at this heritage".[18] Self-evidently, this is not to say that preservation *in situ* excludes any alternative that the preservation of the underwater Cultural Heritage requires under the Convention and in its important Annex.

3. Why is Climate Change Generally not Taken into Account?

With the exception of the 1972 World Heritage Convention, generally Conventions on the protection of Cultural Heritage and property do not expressly take into consideration the climate factor or the more aggressive climate change in today's terminology. *Why* is not an easy question, but deserves to be asked. Follow below some possible explanations.

1. Ordinary annual weather change and climate change vary significantly from country to country. Conventions set out rules that address generally any country regardless of its location on the globe and its degree of exposure to such changes.
2. Such variation in annual weather change can be taken into account by each country when it implements nationally the convention and its protection obligations.
3. Behind the term of Cultural Heritage and property, a variety of materials exist and each is more or less sensitive to climate (see above) and, more recently to climate change; inversely, the conventions set out rules that address generally all the categories of heritage and property that fall under their scope of application.
4. Some conventions, especially the less recent ones, have been negotiated at a time when climate change was not yet identified as a matter both distinct from and more unpredictable than, ordinary annual weather change.

III The Way Forward

It follows that the absence of a clear and explicit mechanism with regard to climate change in the UNESCO conventions *at the time* they were negotiated, as well as in other conventions on the protection of Cultural Heritage and property

---

18 Article 2 (5).

negotiated in the same period of time, can be easily explained. Reasonably, this is not a real *gap* and does not deserve an objective and substantiated criticism. A lot can be done by States Parties to such conventions in implementing them in order to protect Cultural Heritage and property on their territories from the effects of climate change. This is in spite of the absence of such a clear and explicit mechanism in the conventions.

In addition, States Parties do not only adhere to such conventions by implementing them accordingly. It is their commitment towards implementation that may already be considered binding given that these efforts serve as intentional part of a State Party's obligation to perform in good faith. Climate change, needless to say, may affect different parts of a national territory more than others. However, once a State Party adopts a treaty in force it becomes binding upon it and its entire territory.[19]

For future treaty-making undertakings, if and when the international community of States decides that the protection of cultural property and heritage needs to expressly include climate change in international treaties, then the treaties will be drafted differently and openly insert this variable into their mechanisms and provisions.

Until then, in addition to the proper implementation of treaty obligations by each State Party, the international community of States may still take action, if it so wishes, by inserting this *new* variable, climate change, and by dealing with it properly in the Guidelines that States Parties to conventions may establish and that already exist with regard to the Convention concerning the Protection of the World Cultural and Natural Heritage and the Convention for the Safeguarding of the Intangible Cultural Heritage. An alternative option or viable path for the international community of States would be to negotiate *ex novo* another convention in order to include climate change in the protection of Cultural Heritage and property. Regarded realistically this approach appears rather questionable in terms of needs and benefits as well as imponderabilitiy and other constraints.

That option would involve new intergovernmental negotiations and is necessarily time-consuming. It would be faster and more flexible to embed provisions on climate change in the *Guidelines* as if they were existent or anticipated already in advance for a particular convention or at least for a common declaration of understanding among States Parties towards a prospective convention. Again,

19 Vienna Convention on the Law of Treaties; adopted 23 May 1969, entered into force 27 January 1980, 1155 UNTS 331, Articles 26 and 29.

both options are not necessary if States Parties already include climate change in their protection of cultural property and heritage.

There is no doubt that climate change is internationally a serious matter. Being so, it is common sense for States to protect cultural property and heritage in particular, from the new and most devastating threats that climate change represents. These threats add significantly and in unprecedented terms to those related to the ordinary annual climate changes and effect of climate on the natural decay of elements. Ensuring proper mitigation measures for each category of heritage and property, in view of its material and resistance appears more promising and effective for the sake of expedient protection than engaging in a new negotiation for a new convention and waiting for its adoption and entry into force. First, by proper implementation by States Parties, of treaty obligations and of their national existing legislation if adequate from this perspective; second by devising suitable Guidelines or common understandings on existing conventions.

Federico Lenzerini*

Protecting the Tangible, Safeguarding the Intangible: A Same Conventional Model for Different Needs

Abstract The model of the World Heritage Convention, adopted by the UNESCO General Conference in 1972, has emerged as one of the most successful ever established by an international convention aimed at protecting a *common* good, *id est* human rights, Cultural Heritage or environment. At present, the World Heritage Convention has been ratified by 191 states and 1.007 properties are inscribed on the World Heritage List,[1] the keystone around which the system established by the Convention is centred.

In light of the huge success which has characterized the World Heritage Convention, UNESCO has replicated its model in other legal instruments aimed at safeguarding Cultural Heritage in different contexts and/or of different kinds than the World Heritage Convention itself. This has happened, in particular, with respect to the scheme of *enhanced protection* established by the Second Protocol of 1999 to the 1954 Hague Convention on the Protection of Cultural Property in the Event of Armed Conflict as well as to the 2003 Convention for the Safeguarding of the Intangible Cultural Heritage.

As far as the issue of climate change is concerned, both the World Heritage Convention and the Convention for the Safeguarding of the Intangible Cultural Heritage, through providing for general obligations for states parties to take the necessary measures to ensure protection/safeguarding of the heritage they respectively concern, as well as through establishing systems of international cooperation to this end, in principle provide the necessary tools to face the threats posed by climate change on tangible and intangible Cultural Heritage. In practical terms, however, no adequate attention is devoted by both conventions to

---

* Professor of Public International Law and European Union Law at the University of Siena (Italy). Consultant to UNESCO and Legal Advisor of the Italian Ministry of Foreign Affairs at international negotiations concerning the protection of cultural heritage. Member of the Committee on Biotechnology of the International Law Association (ILA) and of the ILA Committee on Cultural Heritage Law, and Rapporteur of the ILA Committee on the Rights of Indigenous Peoples.

1 As of November 2014; for dynamic reference see whc.unesco.org/en/list.

the threats of climate change. Nevertheless, in this respect the World Heritage Convention is certainly better equipped than the Convention for the Safeguarding of the Intangible Cultural Heritage. In any event, the concrete preservation of Cultural Heritage against the above threats ultimately depends on the willingness and capacity of states parties to the respective conventions to take effective and efficient measures in order to properly address the specific problems faced by each manifestation of Cultural Heritage as a consequence of climate change.

I Introduction

The model of the World Heritage Convention (hereinafter *WHC*), adopted by the UNESCO General Conference in 1972,[2] has emerged as one of the most successful ever established by an international convention aimed at protecting a *common* good, *id est* human rights, Cultural Heritage or environment. At present, the WHC has been ratified by 191 states[3] and 1.007 properties are inscribed on the World Heritage List (WHL),[4] the keystone around which the system established by the Convention revolves.

The scope of the WHC is limited to part of the world's immovable cultural property of tangible character and natural sites, namely to those properties which are "of outstanding interest" and therefore "need to be preserved as part of the World Heritage of mankind as a whole".[5] Consequently, adopting a list-based system reflected the logical structure according to which a convention pursuing such a goal should be organized. Although the WHC is organized according to a multi-structured regulation, pursuing the goal of coordinating national[6] and international[7] protection, only the latter is significantly operationalized in the context of the practical implementation of the Convention. In addition, despite the fact that – within the framework of the system of international protection – a specific provision is included stressing that *all* properties of outstanding universal value, and not only those inscribed on the WHL, should be the object of the

2 Convention concerning the Protection of the World Cultural and Natural Heritage; adopted 16 November 1972, entered into force 17 December 1975, 1037 UNTS 151.
3 See UNESCO, 'Legal Instruments' www.unesco.org/eri/la/convention.asp?KO=13055&language=E accessed 24 November 2014.
4 As of November 2014; for dynamic reference confer UNESCO, 'World Heritage List' whc.unesco.org/en/list/accessed 24 November 2014.
5 See WHC, Preamble sixth recital.
6 See Articles 4–5.
7 See Articles 6 ff.

general measures of protection contemplated by the WHC,[8] such a provision is in practice disregarded, attention being exclusively devoted to listed properties.[9] In fact, the listing system represents the aspect making the fortune of the WHC. States parties are eager to inscribe as more national properties as possible on the WHL, due to the huge international visibility a property attains after being listed; as a result of a transitive property, the more the number of national properties inscribed on the List, the greater the degree of international visibility achieved by the state. Also, the willingness to retain such a visibility usually persuades states parties to adopt adequate measures of preservation for the properties concerned, although recently exceptions have occurred leading the World Heritage Committee to delist properties in two cases,[10] giving rise to a potentially dangerous trend for the effectiveness of the WHC. This trend, however, at least for the moment, does not challenge the assumption of the WHC as one of the most successful conventional models ever in the panorama of international law.

II The Right Model for the Wrong Convention

In light of the huge success which has characterized the WHC, UNESCO has replicated its model in other legal instruments aimed at safeguarding Cultural Heritage in different contexts and/or of different kinds than the WHC itself. This has happened, in particular, with respect to the scheme of *enhanced protection* established by the Second Protocol of 1999 to the Hague Convention of 1954 for the Protection of Cultural Property in the Event of Armed Conflict[11] as well as to the 2003 Convention for the Safeguarding of the Intangible Cultural Heritage (CSICH).[12]

The CSICH achieved extensive international support immediately after its adoption. At the moment of this writing, it has been ratified by 161 countries.[13]

---

8 See Article 12.

9 See Federico Lenzerini, 'Article 12. Protection of Properties Not Inscribed on the World Heritage List' in Francesco Francioni and Federico Lenzerini (eds), *The World Heritage Convention. A Commentary* (Oxford University Press 2008) 201. On the functioning of the lists established by Article 11 WHC see Tullio Scovazzi, 'Articles 8–11. World Heritage Committee and World Heritage List' ibid 147; Gionata P Buzzini and Luigi Condorelli, 'Article 11. List of World Heritage in Danger' ibid 175.

10 See notes 24 and 25 below and corresponding text.

11 Adopted 26 March 1999, entered into force 9 March 2004, 2253 UNTS 172.

12 Adopted 17 October 2003, entered into force 20 April 2006, 2368 UNTS 3.

13 See UNESCO, 'Legal Instruments' www.unesco.org/eri/la/convention.asp?language=E&KO=17116 accessed 24 November 2014.

282 elements of intangible Cultural Heritage have been inscribed so far on the Representative List of the Intangible Cultural Heritage of Humanity (established by Article 16 CSICH), 38 on the List of Intangible Cultural Heritage in Need of Urgent Safeguarding (provided for by Article 17 CSICH), and 12 programmes, projects and activities have been inscribed on the Register of Best Safeguarding Practices.[14] In symbolic terms, the adoption of the CSICH determined an epochal step in the context of the process of development of international law on Cultural Heritage; in fact, it epitomized the completion of the evolution of UNESCO standard-setting toward embracing a holistic concept of Cultural Heritage, including not only the material aspects of culture but also the intangible ones.[15] The CSICH, therefore, was quite appealing for non-Western countries, in the context of which the understanding of culture is generally conceived, at least for a large percentage, in its intangible aspects. In fact, since the beginning of its negotiating process, the CSICH was particularly promoted and supported by African, Asian and Latin American countries.[16] Another factor determining the huge success achieved by the CSICH is represented by the circumstance that virtually no real state obligations – in the technical sense of the term – are included in its text. The duties of states parties are in fact expressed in quite soft terms and the Convention looks like a sort of manifesto proclaiming the *right* of states – to be recognized and blessed by the international community – to appropriately safeguard their own intangible Cultural Heritage. Last but not least, a third reason of the success of the CSICH is represented by the perspective that states may use the Convention as a tool for promoting their international visibility, to an equivalent extent of what happens with the WHC; the fact of replicating the model of the latter, based on the listing system, actually makes such a perspective very concrete and appealing.

---

14 See UNESCO, 'Lists of Intangible Cultural Heritage and Register of Best Safeguarding Practices' www.unesco.org/culture/ich/index.php?lg=en&pg=00559 accessed 24 November 2014.

15 Generally on intangible Cultural Heritage and the CSICH see, *inter alia*, Federico Lenzerini, 'Intangible Cultural Heritage: The Living Culture of Peoples' (2011) 22 Eur J Intl L 101; Lucas Lixinski, *Intangible Cultural Heritage in International Law* (Oxford University Press 2013).

16 The very term 'intangible Cultural Heritage' is considered to correspond to a 'loose Eng- lish translation' of the Japanese expression mukei bunkazai; see Richard Kurin, 'Safeguarding Intangible Cultural Heritage: Key Factors in Implementing the 2003 Convention' (2007) 2 International Journal of Intangible Heritage 9, 10.

The CSICH, however, has (or should have) the purpose of safeguarding the common good of intangible Cultural Heritage to the benefit of the international community, rather than a state interest. In this respect, the fact of structuring the operational part of the CSICH according to the model of the WHC appears to be inadequate to ensure appropriate safeguarding for the specificities of intangible Cultural Heritage. The present writer underlined this aspect at the second meeting of experts, preceding the negotiations leading to the adoption of the CSIHC, taking place in Rio de Janeiro in January 2002. He emphasized that 'the schema of the 1972 World Heritage Convention might not be the suitable model for Intangible Cultural Heritage … a legal approach should perhaps avoid the establishment of a List based on selective criteria of importance. The latter might give rise to arbitrary discrimination among cultures'.[17] The 'anti-lists' position was shared by the majority of independent experts, but since the beginning, the intention within UNESCO was to replicate the model of the WHC. This approach easily took the lead at the diplomatic negotiations[18] because it was shared by the most influential state delegations. In fact, it was never seriously challenged, not even when the Norwegian delegate raised the fact that the listing system is not appropriate for intangible Cultural Heritage.[19]

In objective terms, however, the fact remains that the model in point is not the best one – to say the least – in view of appropriately safeguarding intangible Cultural Heritage in light of its specificities and its cultural and spiritual significance, particularly for the communities specifically affected. Indeed, listing intangible Cultural Heritage betrays the inherent value of such a heritage as mirroring the cultural identity of its creators and bearers. In factual terms, the very circumstance of listing inherently presupposes the taxonomy of different manifestations of the heritage concerned; this unavoidably leads to an instinctive perception that the listed examples are particularly valuable and in some way more important than similar manifestations of intangible heritage not included in the same lists. In other words, the establishment of a hierarchy among the different examples of Cultural Heritage ultimately leads to a (mis)understanding – especially among the general public – that certain examples of intangible Cultural Heritage are

17 See UNESCO International Meeting of Experts, 'Intangible Cultural Heritage: Priority Domains for an International Convention' Rio de Janeiro, Brazil (22–24 January 2002) Final Report www.unesco.org/culture/ich/doc/src/00074-EN.pdf accessed 23 September 2013, 8.

18 The present writer was a member of the Italian delegation throughout the whole length of the negotiations leading to the adoption of the CSICH.

19 See Janet Blake, *Commentary on the UNESCO 2003 Convention on the Safeguarding of the Intangible Cultural Heritage* (Institute of Art & Law 2006) 79.

better than others. While this approach can be appropriate – at least partially – for monumental heritage, it is totally improper for immaterial heritage, exactly for the reason that its main significance rests not on its exterior qualities, but rather on the cultural value it has for its creators and bearers. In fact, the main objective of the CSICH, as specified by its Article 1, is 'to ensure respect for the intangible Cultural Heritage of the communities, groups and individuals concerned'; this means that the rationale of the safeguarding of the heritage in point is (*rectius*: should be) grounded on the perception of its significance as an element of the cultural identity of such communities, groups and individuals. Furthermore, one should consider that the action for the safeguarding of intangible Cultural Heritage pursues the goal of preserving the *inherent* value of the said heritage as a vehicle for the protection of cultural diversity. In this respect, it is evident that whatever value judgment based on *external* perceptions determined by its different elements is in principle incompatible with the value of *diversity*. The ultimate danger of the approach in point is represented by the possibility that listing intangible Cultural Heritage may implicitly lead to provoke an instinctive classification of the different communities which create such a heritage, determining an unconscious perception in the public that the communities whose intangible Cultural Heritage is listed are more valuable than others.

Against the reasoning just developed one could object that the Representative List of the Intangible Cultural Heritage of Humanity is contemplated only '(i)n order to ensure better visibility of the intangible Cultural Heritage and awareness of its significance, and to encourage dialogue which respects cultural diversity'; therefore, it would not imply any classification among the different elements of intangible Cultural Heritage based on their quality or value. However, even though this can be true in theory, in practice it is unlikely that the existence of a list will not be perceived by the public as creating a value-based classification among the existing examples of intangible Cultural Heritage, especially among those of a similar kind (e.g. examples of music or theatre representations, which are apparently very similar to each other but, in reality, may be profoundly different on account of the fact that they are part of the cultural identity of *different communities*). The experience of the WHC clearly shows that this conclusion is hardly rebuttable; in fact, although (as previously noted) it includes a provision – Article 12[20] – establishing that the protection accorded by the Convention must not be limited to listed properties, in practice such a provision has remained virtually unapplied and attention has been devoted only to properties

20 See note 8 above and corresponding text.

inscribed on the WHL. Furthermore, it can hardly be asserted that the Representative List of the Intangible Cultural Heritage of Humanity – around which the safeguarding of the intangible Cultural Heritage at the international level is centred – may represent a legal guarantee for the heritage concerned; it rather appears as a tool for states to obtain visibility for the intangible heritage located in their own territory.

III The Common Structure of the WHC and the CSICH and the Threats of Climate Change

It is a fact that the two conventions under debate have many things in common. This holds true not only with respect to their equivalent structure and content, but also for the philosophical rationale inspiring them. Indeed, most probably, the reason why they are almost identical in terms of structure and content is exactly due to their equivalent rationale. While I do not want to question the fact that the initiators of the movement leading to the adoption of the WHC were moved by the noble intention of safeguarding the common good of Cultural Heritage in the general interest of humanity, at the same time one may easily imagine that since the first rounds of negotiations state representatives perceived its enormous potentialities in terms of raising states' international visibility and increasing their attractiveness under different perspectives. The same perception certainly persuaded most countries in the world to support the adoption of the CSICH and to extensively ratify it.[21] The result is two conventions which – although having the huge merit of establishing very successful and efficient legal frameworks for the protection/safeguarding of, respectively, tangible/immovable and intangible Cultural Heritage – pay a huge tribute to the traditional idea of state sovereignty. They in fact put in the hands of states parties the competence/power to determine which elements of heritage are to be protected/safeguarded and the extent to which protection/safeguarding is actually to be accorded. The text and recent practice of the WHC make it crystal clear. As for the former, it is enough to note that no property may be inscribed on the WHL without an explicit request in this respect by the territorial government,[22] as well as that international assistance may be activated by the World Heritage Committee only upon request of the state concerned.[23] With respect to the latter, one

21 See above, text corresponding to note 13.

22 See Article 11(3), stating that '(t)he inclusion of a property in the World Heritage List requires the consent of the State concerned'.

23 See Articles 13(1) and 19.

may consider the cases of the two properties recently delisted from the WHL, namely the Oman's Arabian Oryx Sanctuary[24] and the German Dresden Elbe Valley.[25] The first was delisted in 2007 following a decision of the state party to reduce the area of the property by 90 per cent in order to carry out hydrocarbon prospection in the area. The Dresden Elbe Valley was delisted in 2009 due to the building of a four-lane bridge in the heart of the landscape. In both cases, the decision of the World Heritage Committee was in line with the principles of the WHC, for the reason that the properties concerned had lost their 'outstanding universal value' on account of the modifications occurred with respect to their integrity; in fact, according to Article 1 of the Convention, protection is only granted to cultural properties having such a value (the same is established by Article 2 with respect to natural properties). The point to be emphasized, however, is that these cases show that, when a state party decides not to comply with its obligations pursuant to the WHC, no appropriate manner exists to prevent or react against such behaviour. In fact, it is paradoxical that the only available remedy is to delist the property, because, if the rationale of the Convention is to protect cultural (and natural) properties in the general interest of humanity, then delisting produces the effect of creating a prejudice not for the territorial state, but rather for humanity as a whole. It is, therefore, a sort of *double injury* for humanity, the first being determined by the damage to the integrity of the property determined by the territorial state, the second by the deprivation of significant international protection for it following its delisting. It is even more paradoxical that in the practice the two cases just described have not been treated by other states parties as violations of the obligations established by the WHC, but rather as unfortunate adversities suffered by the territorial states concerned; this is epitomized by the words of the Chair of the World Heritage Committee – María Jesús San Segundo, Ambassador and Permanent Delegate of Spain to UNESCO – following delisting of the Dresden Elbe Valley in 2009: '(e)very time we fail to preserve a site, we share the pain of the State Party'.[26] Such an approach discloses a perception according to which, when a state party has other (usually economic) interests to pursue, Cultural Heritage may be legitimately sacrificed, and that state obtains even the solidarity of its consociates. This gives a very clear idea of

---

24 See UNESCO, 'Arabian Oryx Sanctuary' whc.unesco.org/en/list/654 accessed 5 October 2013.

25 See UNESCO, 'Dresden Elbe Valley' whc.unesco.org/en/list/1156 accessed 5 October 2013.

26 See UNESCO, 'Dresden is deleted from UNESCO's World Heritage List' whc.unesco.org/en/news/522 accessed 5 October 2013.

the manner states conceive their duty to ensure appropriate protection for Cultural and Natural Heritage *in the interest of humanity*... As previously noted, the CSICH imitates the approach and the model of the WHC. For both conventions, state sovereignty is therefore the driving force governing their functioning, in a manner which is evidently at odds with the nature of *common good* of Cultural Heritage of both tangible and intangible character.

In practical terms, the approach just described implies that the two conventions are much more concerned in developing the strategies states may adopt to maximize the potentialities of Cultural Heritage under the perspective of national interests, rather than in properly facing the threats which may jeopardize the integrity and preservation of the heritage concerned to the benefit of future generations. Among the other issues to which such reasoning applies, climate change is undoubtedly included.

Climate change may determine a huge detrimental impact with respect to both tangible and intangible heritage. As regards to the former, the effects of climate change on World Heritage properties – especially mountains, glaciers and sea areas – over the past years are visible to everyone. For example, Mount Kilimanjaro, in Tanzania,[27] has been surrounded by large quantities of ice for some 10,000 years, but in the last decades the amount of ice has drastically decreased, of more than 85 per cent between 1912 and 2011; ice melting on the top of the mountain is at present a constant and seemingly inexorable process, to the point that it is expected to become ice free soon, predictions ranging from 2020 to 2060.[28] Also, in 2004 several properties inscribed on the WHL were affected by the earthquake and tsunami hitting South Asia; cultural properties include the Old Town of Galle and its Fortifications in Sri Lanka[29] as well as Group of Monuments Mahabalipuram[30] and the Sun Temple of Konârak[31] in India; also

---

27 See UNESCO, 'Kilimanjaro National Park' whc.unesco.org/en/list/403 accessed 5 October 2013.

28 See Damien Gayle, 'Kilimanjaro's glaciers shrink and crack as scientists warn Africa's highest mountain may soon be ice free' *Mail Online* (13 November 2010) www.dailymail.co.uk/sciencetech/article-2232195/Kilimanjaros-glaciers-shrink-crack-scientists-warn-Africas-highest-mountain-soon-ice-free.html accessed 5 October 2013.

29 See UNESCO, 'Old Town of Galle and its Fortifications' whc.unesco.org/en/list/451 accessed 5 October 2013.

30 See UNESCO, 'Group of Monuments at Mahabalipuram' whc.unesco.org/en/list/249 accessed 5 October 2013.

31 See UNESCO, 'Sun Temple, Konârak' whc.unesco.org/en/list/246 accessed 5 October 2013.

natural sites were damaged, including the Ujung Kulon National Park[32] and the Tropical Rainforest Heritage of Sumatra,[33] both in Indonesia. Another example is offered by the Everglades National Park,[34] in the United States, which in 2010 was inscribed on the List of World Heritage in Danger, upon request of the American government, because of serious and continuing degradation of its aquatic ecosystem. The Park had been first inscribed on the same list in 1993, due to damage caused by Hurricane Andrew and a marked deterioration in water flows and quality resulting from agricultural and urban development. Although it had been removed from the List in 2007, its degradation has continued, especially in the form of drastic reduction of water inflows and increment of pollution, leading to a loss of marine habitat and decline in marine species.[35] Overall, according to a study of the World Heritage Centre – which is quite outdated at the moment of this writing, as it was carried out in 2005 – 125 properties inscribed on the WHL, located in 59 states parties, were affected at the time by climate change, including *inter alia* coastal marine sites, glaciers, mountainous sites and terrestrial biodiversity sites.[36] The climate threats jeopardizing cultural properties were identified in hurricane, storms, lightening, sea level rise, erosion, flooding, rainfall increase, drought, desertification and rise in temperature.[37] As for the impacts observed for natural properties, they were the following: glacial retreat and glacier melting, sea level rise, loss of biodiversity, species migration and tree-line shift, rainfall pattern changes and occurrence of droughts, frequency of wildfires, coral bleaching, coastal erosion, sea water temperature and salinity change, as well as hurricane, storms and cyclones.[38]

---

32 See UNESCO, 'Ujung Kulon National Park' whc.unesco.org/en/list/608 accessed 5 October 2013.

33 See UNESCO, 'Tropical Rainforest Heritage of Sumatra' whc.unesco.org/en/list/1167 accessed 5 October 2013.

34 See UNESCO, 'Everglades National Park' whc.unesco.org/en/list/76 accessed 5 October 2013.

35 For other examples of world heritage properties actually or potentially affected by climate change see World Heritage Committee, 'Issues Related to the State of Conservation of World Heritage Properties: The Impacts of Climate Change on World Heritage Properties' (Thirtieth Session, Vilnius, Lithuania, 8–16 July 2006) (26 June 2006) WHC- 06/30.COM/7.1, 29–33. See also, in general, UNESCO World Heritage Centre, 'Policy Document on the Impacts of Climate Change on World Heritage Properties' (Paris 2008).

36 Ibid 42–43.

37 Ibid 46.

38 Ibid 44.

Within the WHC framework, climate change is only mentioned in the Operational Guidelines for the Implementation of the World Heritage Convention (Operational Guidelines),[39] inside the Format for the nomination of properties for inscription on the WHL;[40] in particular, at the voice 'environmental pressures', in the context of which the major sources of environmental degradation affecting the property proposed for inscription are to be listed and summarized, climate change is identified as a possible example of such pressures.[41] Also, among the 'selected global conventions and programmes relating to the protection of Cultural and Natural Heritage', the 1992 United Nations Framework Convention on Climate Change[42] is included. However, also in the Convention's text – even though the term *climate change* is not explicitly mentioned – it is dedicated implicit attention. In fact, at Article 11(4), in describing the reasons for which a property already included in the WHL may be inscribed on the List of World Heritage in Danger, certain typical effects of climate change are mentioned, *id est* calamities and cataclysms, landslides, changes in water level, floods and tidal waves. When a World Heritage property is threatened by one of these (or other climate-change-related) phenomena, it is possible to submit a request for international assistance to the World Heritage Committee;[43] the latter may accordingly decide the forms of assistance to be granted to prevent and/or combat the actual or potential effects determined by climate change over the property concerned.

Probably the options offered by the WHC system are still far from laying down a satisfactory set of rules capable of properly addressing the potential threats of climate change on World Heritage properties. However, it is self-evident that in this respect the WHC is certainly better equipped than the CSICH. This is due to at least two reasons. First of all, a very practical one: the heritage object of protection under the WHC is usually much more visible than intangible heritage; therefore, it is much easier to identify, monitor and visualize the effects of climate change – and possibly to keep them under

---

39 The version of the Operational Guidelines updated in 2013 is available at whc.unesco.org/en/guidelines/accessed 5 October 2013.

40 Ibid Annex 5, 102–113.

41 Ibid 109.

42 Adopted 9 May 1992, entered into force 21 March 1994, 1771 UNTS 107.

43 See Articles 19–26 WHC. On international assistance see Anne Lemaistre and Federico Lenzerini, 'Articles 19–26. International Assistance' in Francesco Francioni and Federico Lenzerini (eds), *The World Heritage Convention. A Commentary* (Oxford University Press 2008) 305.

control – on the former than on the latter. Sometimes intangible Cultural Heritage is simply invisible; consequently, it just disappears (this happens in the world on a daily basis), and sometimes nobody realizes that, if not the communities specifically concerned. Secondly, in reproducing the model of the WHC, the CSICH has established a system which, although characterized by equivalent provisions, is much less effective than the WHC in terms of potential action against the effects of climate change; this is due to the different nature – under many perspectives – of the two categories of heritage concerned. In fact, while tangible heritage has a *material structure* on which the necessary measures may be *directly* put into concrete operation – *id est* measures addressing the property *as such* (e.g. restoration, application of layers of protective materials, reinforcement of the structure of a building, cleaning, etc) – this obviously does not apply to intangible heritage, for its ethereal character. When an element of intangible Cultural Heritage is threatened by climate change, clearly the right remedy cannot be that of modifying the nature and characterization of the element concerned; this would in most cases prejudice its very cultural significance for its creators and bearers. Therefore, the only acceptable option is to act on the root causes of climate change with specific respect to the heritage concerned. It is obvious that this is particularly difficult, because intangible Cultural Heritage is by its very nature a *living* heritage, and, since climate change usually forces people to modify their living conditions, habits and traditions, it consequently forces an artificial (nonspontaneous) change in the intangible heritage as well, leading it to lose its distinctive cultural significance, if not to disappear. For example, when a community is forced to leave its ancestral lands for climate-change-related reasons (one may think, for instance, about the so-called *environmental refugees*, *id est* those who are forced to leave an island where they lived which is being submerged due to the increment of the sea level), all elements of intangible Cultural Heritage linked to that specific land will inevitably disappear. Even without any need to refer to such extreme situations, climate change may force communities and even individuals to change their habits so as to adapt them to the new climatic reality, with a consequent loss of intangible Cultural Heritage. Alternatively, it may lead to changes in environmental conditions which are essential for an element of intangible Cultural Heritage to properly fulfil its cultural role as well as to be appropriately preserved and transmitted to future generations. If one takes a look at the two lists established by the CSICH, she/he may easily pick elements of intangible Cultural Heritage which are potentially or actually threatened by climate change. For example, the official Web page of the *Yaokwa, the Enawene Nawe people's ritual for the maintenance of social and cosmic order* (included in the List of Intangible Cultural Heritage in Need of Urgent Safeguarding), in Brazil, describes such a ritual as follows:

> "(t)he Enawene Nawe people … perform the Yaokwa ritual every year during the drought period to honour the Yakairiti spirits, thereby ensuring cosmic and social order for the different clans. The ritual links local biodiversity to a complex, symbolic cosmology that connects the different but inseparable domains of society, culture and nature. It is integrated into their everyday activities over the course of seven months during which the clans alternate responsibilities: one group embarks on fishing expeditions throughout the area while another prepares offerings of rock salt, fish and ritual food for the spirits, and performs music and dance. The ritual combines knowledge of agriculture, food processing, handicrafts (costumes, tools and musical instruments) and the construction of houses and fishing dams. Yaokwa and the local biodiversity it celebrates represent an extremely delicate and fragile ecosystem whose continuity depends directly on its conservation. However, both are now seriously threatened by deforestation and invasive practices, including intensive mining and logging, extensive livestock activity, water pollution, degradation of headwaters, unregulated processes of urban settlement, construction of roads, waterways and dams, drainage and diversion of rivers, burning of forests and illegal fishing and trade in wildlife".[44]

While most of the causes threatening the integrity of the practice in point are man-made (although degradation of headwaters and accidental burning of forests may be effects of climate change), it is based on a fragile ecosystem which may be easily disrupted by climate change. Similar considerations may be developed with respect to the *Traditions and practices associated with the Kayas in the sacred forests of the Mijikenda*, in Kenya, oral traditions and performing arts related to sacred forests, involving the use of natural resources regulated by traditional knowledge and practices, which have contributed to the conservation of local biodiversity.[45] Another example of interest for the present study is offered by the *Sanké mon, collective fishing rite of the Sanké*, in Mali, which was inscribed on the List of Intangible Cultural Heritage in Need of Urgent Safeguarding in 2009 because it is endangered by a number of factors, including degradation of the Sanké lake due to poor rainfall,[46] one of the typical effects of climate change.

---

44 See UNESCO, 'Yaokwa, the Enawene Nawe people's ritual for the maintenance of social and cosmic order' www.unesco.org/culture/ich/index.php?lg=en&pg=00011&USL=00521 accessed 5 October 2013.

45 See UNESCO, 'Traditions and practices associated with the Kayas in the sacred forests of the Mijikenda' www.unesco.org/culture/ich/index.php?lg=en&pg=00011&USL=00313 accessed 5 October 2013.

46 See UNESCO, 'Sanké mon, collective fishing rite of the Sanké' www.unesco.org/culture/ich/index.php?lg=en&pg=00011&USL=00289 accessed 5 October 2013.

The two lists established by the CSICH, however, inescapably offer a very limited picture of the infinite variety of the intangible Cultural Heritage of the world. Many examples of the heritage in point threatened by climate change may therefore be found outside them. It suffices to think about all ice-related traditional practices; for instance, ice-melting in the Arctic is progressively forcing the native Inuit to change their traditional way of life, implying a huge and irreplaceable loss of intangible Cultural Heritage.[47]

Threats to intangible Cultural Heritage determined by climate change are today a daily reality all over the world, and the CSICH does not include any provision dealing with the problem in point. Also, unlike the Operational Guidelines of the WHC, no mention of climate change is included in the Operational Directives for the implementation of the Convention for the Safeguarding of the Intangible Heritage (*Operational Directives*).[48] Certainly the Convention does not prevent the possibility of taking action in view of safeguarding intangible Cultural Heritage against the effects of climate change. This may be done, first of all, through inscribing an element of such a heritage on the List of Intangible Cultural Heritage in Need of Urgent Safeguarding, allowing the element concerned to have *privileged* access to international assistance pursuant to Article 20(a) CSICH. Also, according to Article 11(a) states parties have the duty to 'take the necessary measures to ensure the safeguarding of the intangible Cultural Heritage present in (their) territory', which may well include measures having the purpose of opposing the effects of climate change. Furthermore, the scientific, technical and artistic studies, as well as research methodologies, to be fostered with a view of effectively safeguarding the intangible Cultural Heritage, in particular the intangible Cultural Heritage in danger, pursuant to Article 13(c), may concern action against climate change. Last but not least, action against climate change may be promoted through putting into practice the general duty of international cooperation contemplated by Article 19,[49] by virtue of which states parties 'undertake to cooperate at the bilateral, subregional, regional and international levels' in the general interest of humanity.[50] These, however, are

---

47 See Ed Struzik, 'As Arctic Melts, Inuit Face Tensions with Outside World' *environment360* (1 October 2012), e360.yale.edu/feature/as_arctic_melts_inuit_face_tensions_with_outside_world/2577/accessed 5 October 2013.

48 www.unesco.org/culture/ich/index.php?lg=en&pg=00026 accessed 5 October 2013.

49 Including, *inter alia*, 'the exchange of information and experience, joint initiatives, and the establishment of a mechanism of assistance to States Parties in their efforts to safe-guard the intangible Cultural Heritage'; see Article 19(1).

50 See Article 19(2).

very generic provisions, and their translation into effective measures aimed at addressing the effects of climate change with respect to intangible Cultural Heritage ultimately rests in the hands of states. In other words, for those effects to be properly faced and neutralized, a particularly structured and specific action would be necessary, which is not contemplated by the CSICH, but ultimately depends on the willingness and capacity of states parties to voluntarily take measures in view of properly addressing the specific problems faced by each element of intangible Cultural Heritage as a consequence of climate change, within the broad scope of the freedom left to them by the Convention in safeguarding their national heritage. After all, as is well known, the correct functioning of the CSICH as a whole depends on the willingness of states parties to make it work; in this respect it suffices to think that (like the WHC) for international assistance to be activated, a request from the territorial state concerned is necessary,[51] even with respect to elements inscribed on the List of Intangible Cultural Heritage in Need of Urgent Safeguarding.

In theoretical terms, the interaction between the safeguarding of intangible Cultural Heritage – as determined by the CSICH – and climate change might also be seen under a totally different perspective than the one examined so far. In particular, the hypothesis could be considered that a given element of intangible Cultural Heritage might favour the process of climate change. Should this happen, that particular element would be excluded from the scope of application of the Convention. This is due to the circumstance that, according to its Article 2, '(f)or the purposes of this Convention, consideration will be given solely to such intangible Cultural Heritage as is compatible with existing international human rights instruments, as well as with the requirements of mutual respect among communities, groups and individuals, and of *sustainable development*'.[52] In fact, in the event that one element of intangible Cultural Heritage might produce the effect of favouring climate change, it would clearly be incompatible with the requirements of sustainable development. It is very unlikely, however, that a situation of this kind may actually happen in practice, *id est* that an element of intangible Cultural Heritage may be in itself a force contributing to the progression of climate change.

51 See Article 23 CSICH.

52 Emphasis added.

IV Perspectives in View of Improving Action against Cultural-Heritage-Related Threats Produced by Climate Change

It is a fact that the model of the WHC and the CSICH is not the best possible one in terms of protecting/safeguarding Cultural Heritage. This holds true in particular with respect to the CSICH, since – as seen in the previous section – the WHC system is certainly better equipped than the latter, both in terms of substantive provisions included along its text as well as in *structural* terms, *id est* for the fact that immovable tangible properties offer more chances to organize targeted action against climate change than intangible heritage. This said, to be successful one must work within the reality existing in the real world, finding the right balance between idealism and realism in order to effectively determine positive changes. The current text and structure of the CSICH are carved in stone, and there is no reasonable prospect to change them. At the same time, however, since the Operational Directives are subject to regular revision and may always be modified, they offer a formidable chance to update the global system of the Convention in order to make it more responsive to the need of preventing and combating the effects of climate change on intangible Cultural Heritage. In fact, the Operational Guidelines are an integral part of the CSICH, and the inclusion in their text of specific provisions addressing the issue of climate change would fill the gap left open by the Convention's text in not appropriately dealing with such an issue.

Also, since tangible and intangible heritage often represent two indissoluble components of the same complex cultural reality,[53] promotion of contextual management of the WHC and the CSICH – including in the field of the fight against climate change – might result particularly appropriate and effective.

At the same time, one should not forget that alternative avenues do exist, which, besides the CSICH, establish the conditions for a legal safeguarding of intangible Cultural Heritage. This happens especially with respect to human rights law, the said heritage being an essential part of the cultural identity of communities and individuals. As I have stated in other writings of mine, the recent developments of international law in the field of cultural rights make it reasonable to hold that at present the most effective legal safeguard available for intangible Cultural Heritage is outside the system of the CSICH, in light of the

53 See, in this respect, the third recital of the CSICH Preamble, emphasizing the 'deep-seated interdependence between the intangible Cultural Heritage and the tangible cultural and natural heritage.'

deep interaction between the heritage in point and international human rights.[54] Of course, once a human right exists establishing an international obligation (whereas indirectly) to ensure proper safeguarding of intangible Cultural Heritage of individuals and communities, as part of their cultural identity, such an obligation includes safeguarding the said heritage against the threats of climate change. This argument may be combined – in the form of a mutually-reinforcing relation – with the one concerning the human right to a safe environment, affirmed in contemporary international practice;[55] indeed, the latter right also presupposes a right to be protected against the effects of climate change (although it may be hard to be translated into practice), extended to all essential elements of the life and environment of human beings, including elements of intangible Cultural Heritage.

One final point: in examining the relation between intangible Cultural Heritage and climate change, it is important to emphasize that it may be seen not only under the perspective of the detrimental effects that the latter may determine on the former. On the contrary, it is also opportune to consider that intangible Cultural Heritage may in some cases help in preventing climate change. Many elements of the heritage in point are in fact related to the sustainable use of natural resources, offering lessons which might be used in order to develop a more sustainable model of life, through replacing with long-established uses of traditional communities certain modern habits which contribute to favour climate change.[56]

---

54 See Lenzerini (n 15) 118.

55 See e.g. the practice of the European Court of Human Rights (among others, *López Ostra v. Spain* App no 16798/90 (ECHR, 9 December 1994); *Guerra and Others v. Italy* App no 14967/89 (ECHR, 19 February 1998)) and of the African Commission on Human and Peoples' Rights (in particular, *The Social and Economic Rights Action Center and the Center for Economic and Social Rights v. Nigeria*, Communication no 155/96 (2001) African Human Rights Law Reports 60).

56 See, for example, Benjamin A Gyampoh et al, 'Using traditional knowledge to cope with climate change in rural Ghana' FAO Corporate Document Repository (2001) www.fao.org/docrep/011/i0670e/i0670e14.htm accessed 6 October 2013. The authors, among other things, note that "(t)raditional knowledge – the wisdom, knowledge and practices of indigenous people gained over time through experience and orally passed on from generation to generation – has over the years played a significant part in solving problems, including problems related to climate change and variability. Indigenous people that live close to natural resources often observe the activities around them and are the first to identify and adapt to any changes. The appearance of certain birds, mating of certain animals and flowering of certain plants are all important

V Conclusion

A recent report dealing with the effects of climate change on the life of the Inuit people of Canada notes that their 'millennia-old traditions are already being altered because of the warming Arctic, and (they) face the possibility of having to completely reinvent what it means to be Inuit'.[57] This is an example of a reality, by virtue of which humanity is experiencing an irreplaceable loss of heritage, knowledge and traditions, which ultimately results in a harmful impoverishment of the Cultural Heritage of humankind as a whole. In fact, intangible Cultural Heritage is an essential component of cultural diversity, which, in turn, is a 'common heritage of humanity', as it 'is embodied in the uniqueness and plurality of the identities of the groups and societies making up humankind. As a source of exchange, innovation and creativity, cultural diversity is as necessary for humankind as biodiversity is for nature'.[58] Similar observations may be developed with respect to World Heritage, which, as emphasized by the sixth recital of the WHC Preamble, 'need(s) to be preserved as part of the World Heritage of mankind as a whole'. The international community cannot further postpone setting up a global, targeted and efficient action having the purpose of fighting the detrimental effects produced by climate change on both tangible and intangible Cultural Heritage. Maybe the time is still right to do that; the risk is that very soon it may be too late.

---

signals of changes in time and seasons that are well understood in traditional knowledge systems. Indigenous people have used biodiversity as a buffer against variation, change and catastrophe; in the face of plague, if one crop fails, another will survive [...]. In coping with risk due to excessive or low rainfall, drought and crop failure, some traditional people grow many different crops and varieties with different susceptibility to drought and floods and supplement these by hunting, fishing and gathering wild food plants. The diversity of crops and food resources is often matched by a similar diversity in location of fields, as a safety measure to ensure that in the face of extreme weather some fields will survive to produce harvestable crops".

57 See Unikkaaqatigiit, 'Putting the Human Face on Climate Change – Perspectives from the Inuit in Canada' (2005) www.itk.ca/publication/canadian-inuit-perspectives-climate-change-unikkaaqatigiit accessed 6 October 2013, 11.

58 See UNESCO Universal Declaration on Cultural Diversity, Resolution adopted on the report of Commission IV at the 20th plenary meeting (2 November 2001) Records of the General Conference 31st Session Paris, 15 October to 3 November 2001 Volume 1 Resolutions, Article 1.

Bibliography

Blake, J, *Commentary on the UNESCO 2003 Convention on the Safeguarding of the Intangible Cultural Heritage* (Institute of Art & Law 2006)

Buzzini, GP and L Condorelli, 'Article 11. List of World Heritage in Danger' in Francioni F and F Lenzerini (eds), *The World Heritage Convention. A Commentary* (Oxford University Press 2008) 175

Francioni, F and F Lenzerini (eds), *The World Heritage Convention. A Commentary* (Oxford University Press 2008)

Gayle, D, 'Kilimanjaro's glaciers shrink and crack as scientists warn Africa's highest mountain may soon be ice free' *Mail Online* (13 November 2010) www.dailymail.co.uk/sciencetech/article-2232195/Kilimanjaros-glaciers-shrink-crack-scientists-warn-Africas-highest-mountain-soon-ice-free.html accessed 5 October 2013

Gyampoh, BA, S Amisah, M Idinoba and J Nkem, 'Using traditional knowledge to cope with climate change in rural Ghana' FAO Corporate Document Repository (2001) www.fao.org/docrep/011/i0670e/i0670e14.htm accessed 18 March 2014

Kurin, R, 'Safeguarding Intangible Cultural Heritage: Key Factors in Implementing the 2003 Convention' (2007) 2 International Journal of Intangible Heritage 9

Lemaistre, A and F Lenzerini, 'Articles 19–26. International Assistance' in Francioni F and F Lenzerini (eds), *The World Heritage Convention. A Commentary* (Oxford University Press 2008) 305

Lenzerini, F, 'Article 12. Protection of Properties Not Inscribed on the World Heritage List' in Francioni F and F Lenzerini F (eds), *The World Heritage Convention. A Commentary* (Oxford University Press 2008) 201

– 'Intangible Cultural Heritage: The Living Culture of Peoples' (2011) 22 European Journal of International Law 101

Lixinski, L, *Intangible Cultural Heritage in International Law* (Oxford University Press 2013)

Scovazzi, T, 'Articles 8–11. World Heritage Committee and World Heritage List' in Francioni F and F Lenzerini (eds), *The World Heritage Convention. A Commentary* (Oxford University Press 2008) 147

Struzik, E, 'As Arctic Melts, Inuit Face Tensions with Outside World', *environment360* (1 October 2012) http://e360.yale.edu/feature/as_arctic_melts_inuit_face_tensions_with_outside_world/2577/accessed 5 October 2013

UNESCO, 'Intangible Cultural Heritage: Priority Domains for an International Convention' (International Meeting of Experts, Rio de Janeiro, Brazil, 22–24 January 2002) Final Report www.unesco.org/culture/ich/doc/src/00074-EN.pdf accessed 18 March 2014

UNESCO World Heritage Centre, 'Policy Document on the Impacts of Climate Change on World Heritage Properties' (Paris 2008)

Unikkaaqatigiit, 'Putting the Human Face on Climate Change – Perspectives from the Inuit in Canada' (2005) www.itk.ca/publication/canadian-inuit-perspectives-climate-change-unikkaaqatigiit accessed 18 March 2014

Alessandro Chechi*

The Cultural Dimension of Climate Change: Some Remarks on the Interface between Cultural Heritage and Climate Change Law

Abstract Climate change is one of the major environmental challenges of the twenty-first century whose aggravating effects impact biodiversity, landscapes, the life of peoples and Cultural Heritage. This chapter analyses the different ways climate change threatens Cultural Heritage and examines the existing international legal framework. The aim is to investigate whether, and to what extent, Cultural Heritage law interacts with the international instruments that address the degradation of global climate conditions. The main argument is that these legal regimes are inadequate to protect cultural and environmental resources. Rather than advocating normative or structural reforms, this chapter offers a plea for a new attitude on the part of States, one that aims at sustainable models of economic growth for the sake of the protection of Cultural Heritage of great value to all of humankind.

I Introduction

Climate change is one of the most urgent challenges to humankind and to the sustainability of the world's environment. Despite the uncertainties associated with scenarios and models, it is certain that climate change entails a number of threats: increasing temperatures; change of the frequency, intensity and distribution of extreme events such as droughts, precipitations, floods and tropical cyclones; rising sea level; and increased marine acidification due to rise of carbon dioxide levels in the atmosphere and dissolved in the oceans. There is also no scientific doubt that human activities contribute significantly to climate change.[1]

* Researcher, Art-Law Centre, University of Geneva; PhD in Law, European University Institute (Florence).

1 Solar radiation that reaches Earth warms the surface, which then re-radiates heat back into the atmosphere. Since the atmosphere presents a barrier to both incoming and outgoing radiations, the atmosphere retains a greater proportion of heat radiated from Earth. Human activities perturb this natural filtering (greenhouse) effect by adding greenhouse gases (GHG), such as carbon dioxide and methane, which are the products of fossil fuel combustion resulting from transport-related emissions and

This linkage was uncovered in 2007 by the Intergovernmental Panel on Climate Change (IPCC), which asserted that "there is *very high confidence* that the global average net effect of human activities since 1750 has been one of warming".[2] However, already the 1992 United Nations Framework Convention on Climate Change (UNFCCC) established that climate change can be "attributed directly or indirectly to human activity that alters the composition of the global atmosphere and which is in addition to natural climate variability observed over comparable time periods".[3]

Besides climatic events, human-induced climate change results in global, long-term and accelerating social and economic changes, which threaten human well-being. Further, it has the potential to threaten international peace and security, the stability of nation States[4] and the enjoyment of fundamental human rights. Indeed, climate change will undermine – and is already undermining – the realization of a broad range of individual rights, including the right to health, to life, to food, to water, to shelter, to property, and the rights associated with culture.[5]

While many scholarly works have examined the connection between climate change and internationally protected human rights, this chapter will look at the *cultural dimension* of climate change. In other words, this chapter will focus on the interface between Cultural Heritage law and climate change regulation, addressing the following questions: is Cultural Heritage protection from climate change threats envisaged by international law? Where does international

---

energy production in power plants. The heat-trapping property of these gases leads to an overall warming of the planet and to the ensuing climate change phenomenon. John C Mutter and Kye Mesa Barnard, 'Climate Change, Evolution of Disasters and Inequality', in Stephen Humphreys (ed), *Human Rights and Climate Change* (CUP 2010) 272, 275–277.

2 IPCC, *Fourth Assessment Report: Climate Change* (CUP 2007) 37. The IPCC is the leading international scientific body for the assessment of climate change. It was established by the United Nations Environment Programme and the World Meteorological Organization in 1988 to provide scientific assessments of climate change and its potential environmental and socio-economic impacts, see IPCC, 'Organization' www.ipcc.ch/organization/organization.shtml#.UZJEsUq5-Yk accessed 24 January 2014.

3 Adopted 9 May 1992, entered into force 21 March 1994, 1771 UNTS 107, Article 1.

4 Rosemary Rayfuse and Shirley V Scott, 'Mapping the Impact of Climate Change on International Law', in Rosemary Rayfuse and Shirley V Scott (eds), *International Law in the Era of Climate Change* (Edward Elgar Publishing 2012) 3.

5 Stephen Humphreys, 'Introduction: Human Rights and Climate Change', in Humphreys (n 1) 1.

Cultural Heritage law coincide with or confront obligations under the climate change regime? Where must climate change policies resist with Cultural Heritage law imperatives? Are climate change regime and Cultural Heritage law mutually supportive? This study will thus permit to examine the content and the links between the legal regimes designed by States and international organizations to reduce the vulnerability of Cultural Heritage and the global climate. These resources have emerged as *global public goods*,[6] *id est* values that are fundamental for the international community as a whole and that transcend the interests of individual States.

The 1972 Stockholm Declaration on the Human Environment constitutes the first body of norms and principles governing the collective action of States and other international actors with the goal of safeguarding the essential elements of the environment as a *global public good*, including climate and the biosphere. Subsequently, Resolution 43/53 of the General Assembly underlined that "certain human activities could change global climate patterns, threatening present and future generations"; that "emerging evidence indicates" that the effects of global warming "could be disastrous for mankind if timely steps are not taken at all levels"; and that climate change is a "common concern of humankind".[7] Likewise, the UNFCCC acknowledged that "change in the Earth's climate and its adverse effects are a common concern of humankind".[8] The idea of Cultural Heritage as a 'public good' can be traced back to the Convention for the Protection of Cultural Property in the Event of Armed Conflict, according to which

> "damage to cultural property belonging to any people whatsoever means damage to the cultural heritage of all mankind, since each people makes its contribution to the culture of the world".[9]

---

6 In economic literature, a *global public good* is one that is characterized by non-rivalry, *id est* anyone can use a good without diminishing its availability to others, and nonexcludability, *id est* no one can be excluded from using the good. The term is used in international law discourse to refer to values that are fundamental for the international community as a whole and that transcend the interests of individual States. See Inge Kaul, Isabelle Grunberg and Marc A Stern (eds), *Global Public Goods. International Cooperation in the 21st Century* (OUP 1999); and 'Symposium: Global Public Goods and the Plurality of Legal Orders' (2012) 23 Eur J Intl L 643.

7 UNGA Res 43/53 (6 December 1988) UN Doc A/RES/43/530 Preamble, first, second and third recitals.

8 Preamble, first recital.

9 Adopted 14 May 1954, entered into force 7 August 1956, 249 UNTS, 240 Preamble, second recital.

The same idea was echoed by the Declaration Concerning the Intentional Destruction of Cultural Heritage, which establishes that

> "[...] cultural heritage is an important component of the cultural identity of communities, groups and individuals, and of social cohesion, so that its intentional destruction may have adverse consequences on human dignity and human rights".[10]

This chapter proceeds in five stages. It begins by looking at the ways in which climate change impacts on tangible Cultural Heritage, *id est* art objects, monuments, archaeological sites, landscapes (section II). It then explores Cultural Heritage law (section III) and the international instruments that address the degradation of global climate conditions (section IV) in order to lay the foundation for the examination of their interaction and some of the related problems (section V). The chapter concludes by offering a plea for a new attitude on the part of States, one that meets with the existing international rules and principles and that aims at meaningful sustainable models of economic growth for the sake of the protection of cultural properties of great value to all of humankind as well as future generations (section VI).

II The Impacts of Climate Change on Cultural Heritage

The deleterious effects of climate change will have consequences for the products of human creativity. In the case of built Cultural Heritage, the potential impacts of climate change range from (i) direct physical effects and (ii) effects on social and cultural structures and habitats.[11] This means that the alteration of the climatic equilibrium of the planet does not only severely affect monuments, sites and biodiversity, but has also an impact on people. Therefore, the assessment of the effects of climate change must account for the interactions within and between natural, cultural and societal systems.[12]

---

10 Preamble, fifth recital. The Declaration was adopted on 17 October 2003 by the General Conference of the UNESCO as a reaction to the demolition of the monumental statues of the Buddhas of Bamiyan committed by the Taliban in 2001.

11 UNESCO, 'Case Studies on Climate Change and World Heritage' (World Heritage Centre 2007) whc.unesco.org/en/activities/473 accessed 27 January 2014. This publication presents 26 case studies from selected natural and cultural WHC sites to illustrate the observed and expected impacts of climate change. The World Heritage Centre was established in 1992 and is tasked with managing the day-to-day affairs of the WHC. See whc.unesco.org/en/world-heritage-centre accessed 27 January 2014.

12 UNESCO, 'Report on Predicting and Managing the Effects of Climate Change on World Heritage' whc.unesco.org/en/climatechange accessed 27 January 2014, Paragraph 32.

The adverse impacts of climate change on social and cultural structures can be illustrated by considering the buildings, sites and landscapes where people live, work, worship and socialize. Climate change alters the way people relate to these spaces. For instance, under the pressure of desertification, flooding or sea level rise, populations are forced to migrate, leading to the abandonment of property.[13] This abandonment raises an important concern in contexts where traditional knowledge and skills are essential to ensure a proper maintenance of these properties. Biological changes (with species shifting ranges) can also have an impact on conservation, with the reduction of availability of native species to repair structures and buildings. Moreover, the migration of people leads to the break-up of communities and thus to the eventual loss of social structures, traditional knowledge, cultural identity, rituals and the cultural memory of former inhabitants.[14] For instance, the melting of ice caps and the arctic ground ice as a result of temperature increase in high northern latitudes has already had an impact on the traditional livelihoods and means of survivals of the local indigenous peoples.[15]

Direct physical damages are due to the fact that Cultural Heritage is closely related to the climate. In its Case Studies,[16] the UNESCO World Heritage Centre has distinguished the archaeological heritage from the built heritage in order to identify the most important direct physical effects of climate change.

With respect to the archaeological heritage, the following changes in environmental conditions have been considered: (i) modification in precipitation regimes, regardless of whether the trend marks an increased frequency of droughts or floods or an increased year-to-year variability; (ii) increased soil temperature in response to increased atmospheric temperature, which may provoke subsoil instability, ground heave and subsidence as well as landslides; (iii) sea-level rise,

13 Several islands in the Southern Pacific have already been abandoned after the rising ocean flooded parts of their shorelines and saltwater seeped into the ground water. As 70 per cent of the world population lives in coastal areas, the majority will be affected directly by that development. Stefan Gruber, 'The Impact of Climate Change on Cultural Heritage Sites: Environmental Law and Adaptation' (2008) 8/117 Sidney Law SchoolLegal Studies Research Paper papers.ssrn.com/sol3/papers.cfm?abstract_id=1285741 accessed 26 July 2013, 15–16. See also Hee-Eun Kim, 'Changing Climate, Changing Culture: Adding the Climate Change Dimension to the Protection of Intangible Cultural Heritage' (2011) 18 Intl J Cultural Property 259.

14 UNESCO, 'Case Studies on Climate Change and World Heritage' (n 11) 65.

15 Mutter and Barnard (n 1) 275–276.

16 UNESCO, 'Case Studies on Climate Change and World Heritage' (n 11).

which threatens coastal areas with the subsequent coastal erosion and permanent submersion of low lying areas and the increase in the sea salt chlorides load of coastal soils; (iv) changes in sediment moisture, which are expected to affect data preserved in waterlogged, anaerobic or anoxic conditions; (v) changes in wetting and drying cycles, which will induce crystallization and dissolution of salts and thus affect buried archaeology, wall paintings, frescos, rock art and other decorated surfaces. Essentially, these changes of temperature or water content may jeopardize the conservation of archaeological evidences – even those not known today – and exacerbate decay mechanisms since they perturb the balance in the hydrological, chemical and biological processes of the soil where archaeological sites are located.[17]

Regarding historic buildings the study of the World Heritage Centre evidences that they have a greater intimacy with the ground than modern ones. The built heritage has been designed with the local climate in mind. Historic buildings are more porous and draw water from the ground into their structure and lose it to the environment by surface evaporation. Their wall surfaces and floors are the point of exchange for these reactions. Therefore, increases in soil moisture might result in greater salt mobilisation and consequent damaging crystallisation on decorated surfaces through drying. Moreover, extreme and sudden variations or changes in the amplitude of the diurnal or seasonal variation of temperature and humidity can cause the splitting, cracking, flaking and dusting of materials and surfaces. Timber and other organic building materials may be subject to increased biological infestation such as the migration of pests to altitudes and latitudes that may not have been previously concerned with such threats. Flooding is another important concern. Buildings may be damaged due to the erosive character of rapid flowing water or because they were not designed to withstand prolonged immersion. In addition, post-flood drying may encourage the growth of damaging micro-organisms such as moulds and fungi. Likewise, sea-level rise leads to coastal erosion and threatens coastal properties with total loss.[18]

Another serious threat is desertification. This has been defined as

> [...] land degradation in arid, semi-arid and dry sub-humid areas resulting from various factors, including climatic variations and human activities.[19]

---

17 Ibid 52–53.

18 Ibid 64–65.

19 United Nations Convention to Combat Desertification in those Countries Experiencing Serious Drought and/or Desertification, Particularly in Africa; adopted 14 October 1994, entered into force 26 December 1996, 1954 UNTS 3, Article 1 (a).

Thus, desertification is not only caused by natural causes. Anthropogenic factors such as deforestation, pollution, inappropriate irrigation and poor land governance play a significant part. However, climate change also plays an important role. In the peripheral areas of deserts, sand dunes are normally stabilised by vegetation. Higher temperatures and less rainfall contribute to droughts in such areas, leading to the decline of vegetation and hence to the migration of sand dunes. Once the sand dunes become mobile, all vegetation in their vicinity is at high risk and an unstoppable chain reaction begins. Windstorms, which are increasing in number, extension and frequency due to climate change, further exacerbate the situation.[20]

One of the case studies presented in the report by the World Heritage Centre deals with the three mosques of Djingareyber, Sankoré and Sidi Yahia in Timbuktu, Mali. These sites, which were inscribed in the List of World Heritage in Danger (2012), set up under Article 11 of the UNESCO Convention Concerning the Protection of the World Cultural and Natural Heritage (WHC),[21] bear testimony to the physical and social impacts of climate change (*desertification*). On the one hand, projected changes show that in the future the area of Timbuktu will face a decrease in average rainfall and an increase in atmospheric temperature, which will surely contribute to desert encroachment and sand blown damage.[22] Another climate factor that deserves attention is the increase in extreme precipitation events. Heavy rains in 1999, 2001 and 2003 damaged or caused the collapse of traditional earthen buildings and mud mosques. Moreover, the desertification in the region of Timbuktu is an important source of stress to the three mosques because it might lead the migration of the local population, including the local craftsmen, which are involved in the restoration process of the mud structures of the mosques.[23]

However, there are other climate change-related direct physical impacts that have not been reported by the World Heritage Centre.

First, environmental degradation leads to conditions for the recrudescence of thefts, illicit excavations and exportations of cultural objects. In effect, just as in times of war and of political disorder, the opportunity for illicit trafficking in art increases with the impoverishment of the local population or its departure,

20 Gruber (n 13) 12.

21 Adopted 16 November 1972, entered into force 17 December 1975, 1037 UNTS 151.

22 This threat justified the inscription of Timbuktu on the List of World Heritage in Danger between 1990 and 2005. The site was subsequently re-inscribed in this list in 2012 due to the armed conflict that threatened mosques and tombs.

23 UNESCO, 'Case Studies on Climate Change and World Heritage' (n 11) 74–75.

for example, because of climate change-related events such as desertification and agricultural disruption. As part of the same problem, it appears that this scenario also entails that climate change may affect the efforts currently deployed by national governments, UNESCO and other specialised institutions to enhance (Cultural Heritage) law enforcement in these areas. In addition, climate change induced events may impinge on the implementation of UNESCO's mantras that stolen or illicitly exported cultural materials must be returned to the country of origin and that archaeological assets should be preserved *in situ*.

Second, it may be argued that the geophysical changes resulting from climate change could lead to future armed conflicts, which in turn might impact – as either unwanted or premeditated consequences – cultural sites and properties.[24] The argument about the connection between climate change and conflict boils down to an argument about resource scarcity and competition over the means to sustain livelihoods.[25]

Third, extreme weather events in major art trade hubs like New York and London jeopardize the preservation of movable works of art.[26] These are not unlikely events. In 2012, New York was hit by hurricane Sandy and in 2011 by hurricane Irene, two of the most devastating hurricanes in recent history. These can be related to climate change because, in a warming world, tropical climate conditions expand toward the poles. Consequently, storm tracks that have been associated with the tropics are moving northward (and similar northward shifts will occur in the Pacific). Therefore, areas that previously experienced very few hurricanes and cyclones will start to experience them in greater numbers.[27] Christiane Fischer, President of AXA Art in the Americas observed that

> "Sandy was the costliest event for the art insurance industry by far. *We all will need to adjust our disaster planning strategies to protect important works* [...] *from becoming victims of water damage*".[28]

During hurricane Sandy, Christie's Fine Art Storage Services, a wholly owned subsidiary of the auction house that stores and protects artworks for private and

---

24 Thomas F Homer-Dixon, 'Terror in the Weather Forecast' *New York Times* (New York, 24 April 2007).

25 Ban Ki Moon, 'A Climate Culprit in Darfur' *Washington Post* (Washington, 16 June 2007).

26 On London, see the case study in UNESCO, 'Case Studies on Climate Change and World Heritage' (n 11) 66–69.

27 IPCC (n 2) 46, 53; Mutter and Barnard (n 1) 275–277.

28 AXA ART Group, 'AXA ART Presents Damages from Hurricane Sandy at Its Lounge During 2013 AIPAD Photography Show New York' (press release, 3 April 2013).

institutional owners in Brooklyn, was hit by at least one storm surge. Christie's storage facility is not the only one to have clients' works damaged. The storm threatened manifold businesses, including Chelsea's art galleries, where streets were flooded. It is impossible to know the exact loss as most sources requested anonymity because of the sensitivity of the information. It has been estimated, however, that total losses run in the hundreds of millions.[29]

III Cultural Heritage-Related International Instruments

International Cultural Heritage law has developed throughout time thanks to the standard-setting activity of UNESCO and other international organizations in direct response to the pressure placed upon sites and objects by different forces. The destruction and illicit trade of cultural assets during the two World Wars triggered the adoption of the Convention for the Protection of Cultural Property in the Event of Armed Conflict.[30] The exposure of clandestine excavations and illegal trafficking in cultural objects prompted further international instruments: the Convention on the Means of Prohibiting and Preventing the Illicit Import, Export and Transfer of Ownership of Cultural Property[31] and the Convention on Stolen or Illegally Exported Cultural Objects.[32] Population growth, industrial development and urbanization lead to the adoption of the Recommendation concerning the Preservation of Cultural Property Endangered by Public or Private Works (1968), the WHC, the Recommendation concerning the Protection, at a National Level, of the Cultural and Natural Heritage (1972), and the Recommendation concerning the Safeguarding and Contemporary Role of Historic Areas (1976). As far as the WHC is concerned, it is routinely affirmed that the idea of safeguarding monuments, sites and groups of buildings of 'outstanding universal value' emerged as a result of the international campaign organized in the 1960s by UNESCO to rescue the Abu Simbel Temples in the Upper Nile Valley. These would have flooded following the construction of the Aswan High Dam by the Government of Egypt. Pursuant to a request for assistance from Egypt and Sudan, UNESCO accelerated archaeological research on the site and the temples

29 Laura Gilbert, 'An Exodus from Red Hook' *The Art Newspaper* (London, 26 April 2013).

30 See n 9.

31 Adopted 17 November 1970, entered in force 24 April 1972, 823 UNTS 231.

32 Adopted 24 June 1995, entered into force 1 July 1998, 34 ILM 1322. This treaty was adopted by the International Institute for the Unification of Private Law (UNIDROIT) upon request of UNESCO.

were ultimately dismantled, moved to dry ground and re-assembled. Subsequently, the Convention on the Protection of the Underwater Cultural Heritage[33] was adopted to counter the increasing commercial exploitation of under water Cultural Heritage, which is due to the availability of new technologies that enable discovery and access to previously inaccessible sites.

All in all, these treaties aim to prevent the impoverishment of art-rich countries' patrimony and to guarantee the protection of Cultural Heritage for the sake of its transmission to future generations. However, they were adopted *before* climate change emerged as a harmful phenomenon. This is the reason why climate change is not mentioned in UNESCO agreements. At best, these legal instruments contain references to some of the effects of climate change or to atmospheric pollution. For instance, Article 11(4) WHC states that cultural property can be inscribed in the List of World Heritage in Danger if "threatened by serious and specific dangers, such as the threat of disappearance caused by accelerated deterioration", abandonment, landslides, changes in water level, floods and tidal waves, calamities and cataclysms. Article 6 of the 1978 Recommendation for the Protection of Movable Cultural Property emphasizes that cultural property

> [...] is liable to deterioration as a result of [...] atmospheric pollution [...], which in the long run may have more serious effects than accidental damage or occasional vandalism.[34]

The same holds true for the treaties adopted under the aegis of the Council of Europe. The Convention for the Protection of the Architectural Heritage of Europe affirms that,

> (w)ith a view to limiting the risks of the physical deterioration of the architectural heritage, each Party undertakes: (1) to support scientific research for identifying and analysing the harmful effects of pollution [...]; (2) to take into consideration the special problems of conservation of the architectural heritage in anti-pollution policies.[35]

The Framework Convention on the Value of Cultural Heritage for Society affirms that the States Parties undertake to

---

33 Adopted 2 November 2001, entered into force 2 January 2009, 41 ILM 37.

34 UNESCO, Recommendation for the Protection of Movable Cultural Property, Recommendation adopted on the Report of Programme Commission IV at the thirty-seventh plenary meeting (28 November 1978) Records of the General Conference 20th Session Paris, 24 October to 28 November 1978 Volume 1 Resolutions.

35 Adopted 3 October 1985, entered into force 1 December 1987, CETS No 121, Article 8.

> [...] enrich the processes of economic, political, social and cultural development and land-use planning, resorting to cultural heritage impact assessments [...].[36]

At its 29th session (Durban, 2005), the WHC Committee[37] examined the issue of climate change for the first time. This issue was brought to the attention of the Committee by non-governmental organizations (NGOs) and individuals from several countries. Between 2004 and 2006, these non-State actors had filed five petitions requesting the inscription on the List of World Heritage in Danger for the following WHC sites: Sagarmatha National Park in Nepal; Huascaran National Park in Peru; Great Barrier Reef in Australia; Belize's Barrier Reef Reserve System; and the Waterton-Glacier International Peace Park.[38] Two other petitions were filed in 2007 in regards to the Blue Mountains Area in Australia[39] and La Amistad Park in Panama and Costa Rica.[40] Lastly, in 2009, two NGOs filed a

36 Adopted 27 October 2005, entered into force 1 June 2011, CETS No 199, Article 8.

37 In the machinery of the WHC, the Committee consists of representatives from 21 of the States Parties to the Convention, which are elected for terms up to six years by the General Conference. It is tasked with: (i) inscribing in the WHC List the sites of 'outstanding universal value' designated by the States Parties; (ii) monitoring the state of conservation of WHC properties; (iii) establishing the terms for use of the WHC Fund; (iv) allocating financial assistance upon requests from States Parties; and (v) establishing a List of World Heritage in Danger. See Articles 8–14 WHC.

38 The WHC does not expressly authorize petitions of this nature by NGOs or individuals. The petitioner on the Huascaran National Park case relied on a UNESCO document (UNESCO, 'World Heritage Information Kit' (World Heritage Centre 2008) whc.unesco.org/documents/publi_infokit_en.pdf accessed 27 January 2014). It indicates that "private individuals, non-governmental organizations, or other groups may also draw the Committee's attention to existing threats. If the alert is justified and the problem serious enough, the Committee may consider including the site on the List of World Heritage in Danger" (18). The other petitions simply cited the power of the WHC Committee to add sites to the List of World Heritage in Danger as a justification. For an analysis of these petitions see Erica J Thorson, 'The World Heritage Convention and Climate Change: The Case for a Climate-Change Mitigation Strategy beyond the Kyoto Protocol' in William C G Burns and Hari M Osofsky (eds), *Adjudicating Climate Change: State, National, and International Approaches* (CUP 2009) 255.

39 See Climate Justice, 'International Tribunals' www.climatelaw.org/cases/country/intl accessed 27 January 2014.

40 See International Environmental Law Project of Lewis and Clark Law School, 'Petition to the World Heritage Committee requesting inclusion of Talamanca Range-La Amistad Reseres/La Amistad National Park on the List of World Heritage in Danger' law.lclark. edu/live/files/195 accessed 27 January 2014.

petition focusing on the threat posed by black carbon to an array of WHC sites.[41] These petitions argued that, pursuant to their obligations under the WHC, States Parties should make drastic cuts in their GHG emissions, regardless of their commitments under the UNFCCC and the Kyoto Protocol.[42]

With Decision 29 COM 7B.a of 2005, the WHC Committee took note of the petitions and of the 'genuine concern' of the petitioners and acknowledged

> "that the impacts of climate change are affecting many and are likely to affect many more World Heritage properties [...] in the years to come".[43]

It then encouraged States Parties to incorporate responses to these threats in management plans developed for WHC sites. Furthermore, it requested the World Heritage Centre, in collaboration with the WHC's Advisory Bodies,[44] interested States Parties and the petitioners, to convene a working group of experts to

> "review the nature and scale of the risks posed to World Heritage properties arising specifically from climate change; [...] jointly develop a strategy to assist States Parties to implement appropriate management responses [...]"[45]

and to

> "prepare a joint report on 'Predicting and managing the effects of Climate Change on World Heritage' to be examined by the Committee at its 30th session (Vilnius, 2006)".[46]

As a result, the Committee did not inscribe the sites in question in the List of World Heritage in Danger as requested by the petitioners.[47]

The expert meeting took place in March 2006 at the UNESCO headquarters in Paris and resulted in the preparation of a Report on Predicting and Managing the Effects of Climate Change on World Heritage ('the Report')[48] and a Strategy

---

41 See Earthjustice and ACJP, 'Petition to the World Heritage Committee' whc.unesco.org/uploads/activities/documents/activity-393-4.pdf accessed 27 January 2014.

42 Adopted 11 December 1997, entered into force 16 February 2005, 2303 UNTS 148.

43 UNESCO, World Heritage Committee Decision 29COM 7B.a (9 September 2005) WHC-05/29.COM/22, Paragraphs 3–5.

44 These are the International Council for Monuments and Sites (ICOMOS), the International Union for Conservation of Nature (IUCN) and International Centre for the Study of the Preservation and Restoration of Cultural Property (ICCROM).

45 UNESCO (n 43) Paragraph 7.

46 Ibid Paragraph 9.

47 For an assessment of the Committee's decision see William C G Burns, 'Belt and Suspenders? The World Heritage Convention's Role in Confronting Climate Change' (2009) Rev Eur Community and Intl Environmental L 148.

48 See n 12.

to Assist State Parties to Implement Appropriate Management Responses ('the Strategy').[49] The Report (i) provided a detailed assessment of the potential – direct, social and cultural – impacts of climate change on World Heritage cultural sites; (ii) required the adoption of management strategies for site-based mitigation and adaptation responses; and (iii) considered the development of synergies and partnerships with other international treaties, such as the UNFCCC and the Kyoto Protocol, the Ramsar Convention on Wetlands,[50] the UNESCO's Programme on Man and the Biosphere and the Convention on Biological Diversity.[51] The Strategy focused on three types of actions to safeguard WHC properties: (i) preventive actions, including monitoring, reporting and mitigation of climate change impacts; (ii) corrective actions, with a focus on global, regional and local adaptation strategies; and (iii) the sharing of knowledge, including best practices, education and capacity building.

The WHC Committee endorsed these two documents at its 30th session (Vilnius, 2006) and called on all the States Parties to implement the Strategy so as to protect the outstanding universal values, integrity and authenticity of WHC properties from the adverse impacts of climate change.[52] The Committee further requested the World Heritage Centre to develop a policy document on the impacts of climate change on WHC properties. The Committee stated that this document should focus on: (i) synergies between conventions; (ii) identification of future research needs in this area; (iii) legal questions on the role of the WHC with regard to appropriate responses to climate change; (iv) linkages to other UN and international bodies dealing with the issues of climate change; (v) alternative mechanisms, other than the List of World Heritage in Danger, to address climatic change and other international concerns.[53] The Policy Document on the Impacts of Climate Change on World Heritage Properties[54] was presented and discussed by the WHC Committee at its 31st session (Christchurch, 2007) and

49 UNESCO, 'Strategy to Assist States Parties to Implement Appropriate Management Responses' (26 June 2006) WHC-06/30.COM/7.1.

50 Convention on Wetlands of International Importance especially as Waterfowl Habitat; adopted 2 February 1971, entered into force 21 December 1975, 996 UNTS 245.

51 Adopted 5 June 1992, entered into force 29 December 1993, 1760 UNTS 79.

52 UNESCO, World Heritage Committee Decision 30 COM 7.1 (23 August 2006) WHC06/30.COM/19, Paragraphs 6–8.

53 Ibid Paragraph 13.

54 UNESCO World Heritage Centre, 'Policy Document on the Impacts of Climate Change on World Heritage Properties' (Paris 2008) whc.unesco.org/uploads/activities/documents/activity-393-2.pdf accessed 27 January 2014.

subsequently approved by the 16th General Assembly of the States Parties in 2007, which also encouraged UNESCO and the WHC Advisory Bodies to disseminate and promote the application of the Policy Document, the Report and the Strategy, including to the general public.

At its 32nd session (Quebec City, 2008) the WHC Committee decided to adopt "criteria for assessing properties which are most threatened by climate change for inclusion on the List of World Heritage in Danger".[55] This led to the amendment of the Operational Guidelines.[56] Now, Paragraph 179(b)(vi) of the Operational Guidelines establishes that a WHC site can be inscribed on the List of World Heritage in Danger when the Committee finds that the property is faced with a potential threat which could have deleterious effects on its inherent characteristics, such as "threatening impacts of climatic, geological or other environmental factors".[57] New Paragraph 181 reads:

> "(T)he threats and/or their detrimental impacts on the integrity of the property must be those which are amenable to correction by human action. In the case of cultural properties, both natural factors and man-made factors may be threatening, while in the case of natural properties, most threats will be man-made and only very rarely a natural factor (such as an epidemic disease) will threaten the integrity of the property. In some cases, the threats and/or their detrimental impacts on the integrity of the property may be corrected by administrative or legislative action, such as the cancelling of a major public works project or the improvement of legal status".

The amendment of the Operational Guidelines is important in that it permits to emphasise that the legal, factual and political context in which the provisions of an international treaty are applied may have changed as compared to the circumstances prevailing at the time of its adoption. In the case of the WHC, which has been in existence for more than forty years, the correct implementation of its norms has not required the revision of its text.[58] Rather, the WHC Committee has taken into account the evolving context of natural and Cultural Heritage and new developments in international law through the continuing revision of the Operational Guidelines.[59] An example is provided by the addition of the category

---

55 UNESCO, World Heritage Committee Decision 32 COM 7A.32 (31 March 2009) WHC08/32.COM/24Rev para 5.

56 UNESCO, 'Operational Guidelines for the Implementation of the World Heritage Convention' (June 2013) WHC. 13/01 whc.unesco.org/en/guidelines accessed 27 January 2014.

57 Paragraph 180(b)(v) lists the same potential danger for natural properties.

58 Article 37 WHC regulates the issue of the revision.

59 Francesco Francioni, 'Thirty Years On: Is the World Heritage Convention Ready for the 21st Century?' (2003) Italian Y Intl L 13, 28.

cultural landscape to the heritage defined in Articles 1 and 2 WHC. Another example is represented by the addition of a new criterion for listing natural sites to take into consideration the value of biodiversity following the adoption of the Convention on Biological Diversity.[60] Accordingly, the Operational Guidelines may represent, in accordance with Article 31(3) of the Vienna Convention on the Law of Treaties (VCLT),[61] a "subsequent agreement between the parties regarding the interpretation of the treaty or the application of its provisions" (letter a) or "subsequent practice in the application of the treaty which establishes the agreement of the parties regarding its interpretation" (letter b).[62]

In light of the foregoing survey, it appears that the WHC Committee's response to the impacts of climate change on WHC properties is quite disappointing. In effect, although it acknowledged that numerous sites are threatened by climate change,[63] it opted for a soft approach motivated by the alleged primary role of the IPCC and of the UNFCCC in addressing the problem of climate change at the international and national level.[64] Accordingly, beyond the monitoring of sites pursuant to Article 29 WHC, the Committee now has only two options to react to a threat to or possible destruction of WHC properties: the inscription on the List of World Heritage Sites in Danger or the deletion of the site in question when it loses its outstanding universal value.[65] The listing may not only attract interna-

60 Amendment of the Operational Guidelines approved by the WHC Committee at its 16th session (Santa Fe, 1992) (14 December 1992) WHC-92/CONF.002/12.

61 Adopted 23 May 1969, entered into force 27 January 1980, 1155 UNTS 331.

62 Gionata P Buzzini and Luigi Condorelli, 'Article 11: List of World Heritage Sites in Danger and Deletion of a Property from the World Heritage List', in Francesco Francioni (edited with the assistance of Federico Lenzerini), *The 1972 World Heritage Convention. A Commentary* (OUP 2008) 175, 189.

63 See Decision 29 COM 7B.a (n 43) Paragraph 5; and the survey in the Report (n 12) Paragraphs 39–46, which indicates that of "the 110 responses received from 83 States Parties, 72 per cent acknowledged that climate change had an impact on their natural and Cultural Heritage", and that a "total of 125 World Heritage sites were mentioned specifically as threatened by Climate Change".

64 The Strategy (n 49) states that the "UNFCCC is the UN instrument through which mitigation strategies at the global and States Parties level is being addressed" (Paragraph 18).

65 As of July 2013, the WHC Committee has delisted two sites only: the Dresden Elbe Valley in 2009 (Decision 33COM 7A.26) and the Arabian Oryx Sanctuary in 2007 (Decision 31COM 7B.11). On the List of World Heritage Sites in Danger see Buzzini and Condorelli (n 62); on the Fund, see Federico Lenzerini, 'Articles 15–16: World Heritage Fund' in Francioni and Lenzerini (n 62) 269.

tional assistance, but it also allows the Committee to assign financial resources from the Fund established under Article 15 WHC.[66] Moreover, the listing and the threat of deletion represent powerful tools to discourage States Parties from jeopardising the outstanding universal value of their properties.[67]

However, although these options are very strong, in-danger listing and deletion do not seem adequate to pursue the objectives set forth in Cultural Heritage treaties, *id est* the conservation and protection of the world's heritage for the sake of its transmission to future generations. Stated differently, it appears that these measures cannot effectively prevent the deterioration or disappearance of sites of outstanding universal value.

Apart from the fact that the limited size of the Fund does not allow major or multiple investments regarding properties endangered by climate change, it is worth considering that the deterrent effect of the listing would diminish if it were more common. In other words, an increase in in-danger listings related to climate change threats by the WHC Committee would likely weaken the credibility of this measure.[68]

IV Climate Change-Related International Regime

The centrepiece of global efforts to combat climate change is the UNFCCC. It was signed in 1992 at the United Nations Conference on Environment and Development in Rio de Janeiro – together with the Convention to Combat Desertification and the Convention on Biological Diversity – as global warming had become a priority on the international political agenda. The UNFCCC defines climate change as

> […] a change of climate which is attributed directly or indirectly to human activity that alters the composition of the global atmosphere and which is in addition to natural climate variability observed over comparable time periods.[69]

---

66 Paragraph 189 of the Operational Guidelines (n 56) provides that the "Committee shall allocate a specific, significant portion of the World Heritage Fund to financing of possible assistance to World Heritage properties inscribed on the List of World Heritage in Danger".

67 Gruber (n 13) 9.

68 Gruber (n 13) 9. As of July 2013, 44 properties are inscribed on the List of World Heritage in Danger, whereas 1.007 are inscribed on the World Heritage List.

69 Article 1.

Therefore, the Convention distinguishes between anthropogenic climate change and the variability of climatic conditions attributable to natural causes. The ultimate objective of the UNFCCC is to achieve

> (S)tabilization of greenhouse gas concentrations in the atmosphere at a level that would prevent dangerous anthropogenic interference with the climate system [...] within a time-frame sufficient to allow ecosystems to adapt naturally to climate change, to ensure that food production is not threatened and to enable economic development to proceed in a sustainable manner.[70]

The Convention also sets out some guiding principles. In particular, it refers to the precautionary principle – according to which the lack of scientific certainty should not be used as an excuse to postpone action when there is a threat of serious or irreversible damage – and to the principle of common but differentiated responsibility – which is based on the facts that developed and developing States contributed differently to global environmental problems and that some States have less ability to cope with environmental problems.[71]

According to Article 4(1) of the UNFCCC, States Parties are required to: (i) develop, periodically update, publish and make available to the Conference of the Parties national inventories of anthropogenic emissions by sources and removals by sinks of all greenhouse gases; (ii) formulate, implement, publish and regularly update national and, where appropriate, regional programmes for mitigating climate change; (iii) promote and cooperate in the development and diffusion of technologies, practices and processes that control, reduce or prevent anthropogenic emissions of greenhouse gases; (iv) develop strategies for adapting to the impacts of climate change; (v) promote the sustainable management of resources; (vi) enhance GHG sinks and reservoirs (such as forests); and (vii) take climate change into account in their relevant social, economic, and environmental policies and cooperating in scientific, technical, and educational matters, as well as public awareness. Moreover, the UNFCCC stipulates that industrialized countries should support climate change activities in developing countries by providing financial assistance and sharing technologies. The Convention's objective is to help developing countries to limit GHG emissions in ways that do not hinder their economic development. The Conference of the Parties uses the

---

70 Article 2.

71 Article 3 UNFCCC states that "on the basis of equity and in accordance with their common but differentiated responsibilities and respective capabilities [...] the developed country Parties should take the lead in combating climate change and the adverse effects thereof".

information provided through the national reports to assess and review the effective implementation of the UNFCCC and to evaluate the aggregated effect of the steps taken by States Parties.

The UNFCCC has achieved almost universal participation.[72] However, it is a binding treaty with no binding obligations. One reason is that the UNFCCC was concluded in a moment where there existed uncertainties on the phenomena of global warming and climate change. Another reason is that various States – most notably the United States – refused to accept legally binding reduction targets for GHG emissions. Consequently, the UNFCCC merely calls on the (industrialized) States Parties to 'aim' to return their GHG emissions back to 1990 levels without setting precise targets.[73] In sum, developed States agreed to voluntary caps on emission of man-made GHG.[74]

However, by 1995, the GHG emissions of most developed States were already well above 1990 levels. The realization that more substantive measures were necessary to confront global warming and the other dangerous effects of climate change led to the adoption of the Kyoto Protocol to the UNFCCC at the Third Conference of the Parties in 1997. Essentially, the Protocol was introduced to expand the scope of the UNFCCC in order to achieve an ambitious reduction – rather than stabilization – of GHG emissions.[75]

The Kyoto Protocol sets up a regulatory regime providing concrete and specific binding obligations: it called 37 industrialized States and the European Union to "reduc(e) their overall emissions of such gases by at least 5 per cent below 1990 levels in the commitment period 2008 to 2012".[76] In addition, the Protocol required States Parties to begin negotiating commitments for subsequent periods by 2005.[77] In line with the UNFCCC and in accordance with the principle of common but differentiated responsibilities, developing countries were not required to

72 As of July 2013, it has been ratified by 194 States and 1 regional economic integration organization, the European Union.

73 Article 4(2)(b).

74 Daniel Bodansky, 'The United Nations Framework Convention on Climate Change: A Commentary' (1993) Yale J Intl L 451.

75 See UNFCCC, 'Making Those First Steps Count: An Introduction to the Kyoto Protocol' unfccc.int/essential_background/kyoto_protocol/items/6034.php accessed 27 January 2014.

76 Article 3(1).

77 Article 3(9). The second commitment period (2013–2020) was decided on 8 December 2012, with the adoption of the Doha Amendment to the Kyoto Protocol. See UNFCCC, 'Doha Amendment' unfccc.int/kyoto_protocol/doha_amendment/items/7362.php accessed 27 January 2014.

make any commitment because it was considered that the largest responsibility for the levels of GHG emissions in the atmosphere laid with industrialized countries. Moreover, similarly to the UNFCCC, the Kyoto Protocol contains rules providing for concessions to wavering States, which substantially dilute the Parties' commitments. States' actual emissions have to be monitored. Therefore, States are required to submit reports about annual emission inventories.

Under the Kyoto Protocol, States must meet their reduction targets for GHG emissions primarily through national measures. In addition to these means, the Protocol introduced three market-based mechanisms: (1) Joint Fulfilment of Commitments and Implementation (Joint Implementation); (2) the Clean Development Mechanism (CDM); (3) Emissions Trading.[78]

The mechanism of Joint Implementation[79] allows any Party included in Annex I (States with an emission reduction or limitation commitment under the Kyoto Protocol) to transfer to, or acquire from, any other Annex I Party Emission Reduction Units (ERUs) resulting from projects aimed at reducing anthro pogenic emissions by sources or enhancing anthropogenic removals by sinks of GHG in any sector of the economy. Therefore, Joint Implementation projects provide a reduction in emissions by sources or an enhancement of removals by sinks that is additional to what would otherwise have occurred. One of the purposes of Joint Implementation projects is to involve private-sector money in the transfer of technology and know-how.

The CDM[80] allows any Party included in Annex I (States with an emission reduction or limitation commitment under the Kyoto Protocol) to implement an emission-reduction project in developing countries. Such projects can earn saleable Certified Emission Reduction (CER) credits that can be counted towards meeting Kyoto targets. The projects must qualify through a rigorous and public registration process. Approval is given by the Designated National Authorities.

Emissions trading[81] is a mechanism whereby any Party included in Annex I (States with an emission reduction or limitation commitment under the Kyoto Protocol) may trade their emission allowances with other States Parties. These emissions allowances correspond to emission units permitted but not used by States.

---

78 See UNFCCC, 'The Mechanisms under the Kyoto Protocol' unfccc.int/kyoto_protocol/mechanisms/items/1673.php accessed 27 January 2014. For an analysis of these mechanisms see Rafael Leal-Arcas, *Climate Change and International Trade* (Edward Elgar Publishing 2013) 227 ff.

79 Article 6 Kyoto Protocol.

80 Article 12 Kyoto Protocol.

81 Article 17 Kyoto Protocol.

Thus, the emissions trading mechanism created a new commodity. However, since carbon dioxide is the principal GHG, people speak of carbon trading rather than emission reductions. Carbon is now traded like any other commodity. Apart from actual emissions units, other units can be traded under the Kyoto Protocol's emissions trading scheme: 'removal units' on the basis of land use, land-use change and forestry activities such as reforestation; 'emission reduction units' generated by a joint implementation project; certified emission reductions generated from a clean development mechanism project activity. Transfers and acquisitions of these units are tracked and recorded through the registry systems under the Kyoto Protocol.

These market-based mechanisms are meant to: (i) enable industrialized States Parties to meet their emission targets in a cost-effective way; (ii) to stimulate technology transfer and 'green' investments; (iii) encourage the private sector and developing countries to contribute to emission reduction efforts.

All in all, the UNFCCC and the Kyoto Protocol require State action. In this respect it is necessary to distinguish between adaptation and mitigation. Adaptation refers to "adjustments in ecological, social, or economic systems in response to actual or expected climatic stimuli and their effects or impacts. It refers to changes in processes, practices, and structures to moderate potential damages or to benefit from opportunities associated with climate change".[82] Therefore, it focuses on reducing vulnerability to the impacts of climate change.

Solutions to adapt to the effects of climate change take many shapes and forms depending on the context and the specific impacts of climate change. For instance, it can range from building flood defences, setting up early warning systems for cyclones and switching to crops that fare better under drought conditions to redesigning communication systems, business operations and government policies.[83] On the other hand, mitigation corresponds to any "anthropogenic intervention to reduce the sources or enhance the sinks of greenhouse gases".[84] Therefore, it refers to actions that lead to reducing the emissions of GHGs.[85]

---

82 IPCC, 'Third Assessment Report, Synthesis Report, Working Group II: Impacts, Adaptation, and Vulnerability (Chapter 18.2)' (2001) www.grida.no/publications/other/ipcc_tar/?src=/climate/ipcc_tar/wg2/644.htm accessed 27 July 2013.

83 See UNFCCC, 'Focus: Adaptation' unfccc.int/focus/adaptation/items/6999.php accessed 27 January 2014.

84 IPCC, 'Third Assessment Report, Synthesis Report, Working Group III: Mitigation (Appendix II: Glossary)' (2001) www.grida.no/publications/other/ipcc_tar/?src=/climate/ipcc_tar/wg3/466.htm accessed 27 July 2013.

85 See UNFCCC, 'Focus: Mitigation' unfccc.int/focus/mitigation/items/7169.php accessed 27 January 2014.

V What Relationship between Cultural Heritage and Climate Change?

Having explored the legal regimes for the protection of Cultural Heritage and the global climate, it is now necessary to examine the interface between these areas of international law. The main goal of this analysis is to discuss whether these legal systems should become mutually supportive at both the legal and policy levels.

However, the analysis that follows will focus only on the WHC. This is for three main reasons: the issue of climate change has been examined and internalized by the WHC Committee; the direct, social and cultural effects on several WHC sites have been vastly documented; it is critical to the survival of many items of Cultural Heritage that the WHC works closely with the climate change regime. At this juncture, it must also be emphasized that this convention applies to heritage as defined in Articles 1 and 2 situated on the territory of States Parties, whether or not this heritage has been inscribed on the WHC List under Article 11.[86]

1. The Story So Far

The foregoing examination reveals that climate change law neither addresses the impacts of climate change on Cultural Heritage nor contains references to UNESCO treaties. For instance, there is no Kyoto Protocol's CDM projects directly involving Cultural Heritage.[87] One could find various explications to this lacuna. First, it can be argued that in 1992, when the UNFCCC was concluded, the scientific community had not yet realized what consequences could arise from the changing climate. Second, it can be submitted that the protection of Cultural Heritage is not reflected in climate change instruments because it has been too difficult for environmental negotiators to include this issue into the already thorny climate change negotiations. Third, it can be argued that in the face of catastrophic predictions regarding climate change, the less appalling consequences regarding individual sites have been overlooked.

By way of contrast, although the WHC does not mention climate change specifically, the WHC Committee has underlined the key role of the IPCC, the

86 Guido Carducci, 'Articles 4–7: National and International Protection of the Cultural and Natural Heritage' in Francioni and Lenzerini (n 62) 103, 113.

87 See UNFCCC, 'Project Search' cdm.unfccc.int/Projects/projsearch.html accessed 27 January 2014.

UNFCCC and the Kyoto Protocol in various documents. The Report affirms that "mitigation at the global and States Parties level is the mandate of the UNFCCC and its Kyoto Protocol"[88] and that the WHC Committee "could collaborate with the UNFCCC secretariat on Climate Change issues by presenting information at the Conference of the Parties (COP) and subsidiary bodies meetings [...]".[89] Moreover, the Strategy states that the "UNFCCC is the UN instrument through which mitigation strategies at the global and States Parties level is being addressed" and that the "World Heritage community could participate in climate change mitigation" by providing "information to IPCC and UNFCCC on the impacts of climate change" on WHC sites and by assisting them in tailoring mitigation strategies and in encouraging site-based reduction of GHG emissions.[90] The Policy Document calls for increased cooperation between WHC bodies and other international conventions and organizations working on climate change. For instance, it states that the World Heritage Centre and the Advisory Bodies "will seek to take advantage of synergies to better coordinate and enhance effective implementation" of the WHC "by capitalizing upon each organization's strengths, and aiming to avoid overlap and duplication with, and respect the individual mandates of, other international organizations and mechanisms"; and that the "World Heritage Centre will strengthen its relationship with the UNFCCC and IPCC Secretariats, which are the key international organizations working on climate change".[91] Moreover, the Policy Document establishes that climate change should be considered in the application of the WHC and the Operational Guidelines, *id est* in the preparation of nominations, in the drafting and implementation of the management plans by States Parties, in the monitoring by the WHC Committee and in the reporting by States Parties.[92] Finally, the Operational Guidelines call for "appropriate co-ordination and information-sharing between the World Heritage Convention and other Conventions, programmes and international organizations related to the conservation of Cultural and natural heritage", including the UNFCCC.[93]

In sum, it appears that Cultural Heritage law and climate change regulation are not mutually supportive. To use a metaphor, it is as if, within the UN family,

88 UNESCO, 'Report' (n 12) Paragraph 6; see also Paragraphs 7, 77, 121, 124–125.
89 Ibid Paragraph 61.
90 UNESCO, 'Strategy' (n 49) Paragraphs 17–19.
91 UNESCO, 'Policy Document' (n 54) 4.
92 Ibid 6–8.
93 UNESCO, 'Operational Guidelines' (n 56) Paragraphs 42, 44.

one of the sisters is so absorbed in her own pursuits to the point of ignoring one of her sisters and their commonalities.

2. Synergies between Treaty Regimes

Various questions arise as a result of the existing state of affairs: does it make sense to argue for the establishment of reciprocal linkages between Cultural Heritage and climate change law? Would a more coordinated international legal response permit to enhance the protection of these two fundamental *global public goods*? What are the most reasonable solutions to build synergies between these two areas of international law?

The main reason in favour of the establishment of synergies between Cultural Heritage and climate change law is that this could bring about various advantages. This is due to the fact that State obligations under these regimes appear to be complementary:[94] while the main goals of climate change regulation are about environmental protection, sustainable development and the preservation of ecosystems for present and future generations, the main objective of the Cultural Heritage regime is the preservation of all tangible and intangible manifestations of culture having artistic, historical and symbolic values, and their transmission to future generations.[95] Today, cultural assets can be seen as part of the environment where they have been created and as an essential component of the identity and history of the people who created them or for whom they were created. This role of Cultural Heritage as part of public space opens the way to a holistic approach to heritage, *id est* an approach that brings together Cultural and Natural Heritage.[96]

The Policy Document cites some key advantages. First, it emphasises that WHC properties could be used "as a means to raise awareness about the impacts of climate change upon World Heritage to act as a catalyst in the international

94 Humphreys (n 5) 11.

95 On this commonality see Article 1 of the 2001 Universal Declaration on Cultural Diversity, which famously states that "[…] cultural diversity is as necessary for humankind as biodiversity is for nature. In this sense, it is the common heritage of humanity and should be recognized and affirmed for the benefit of present and future generations". Resolution adopted on the report of Commission IV at the 20th plenary meeting (2 November 2001) Records of the General Conference 31st Session Paris, 15 October to 3 November 2001 Volume 1 Resolutions.

96 Francesco Francioni, 'Public and Private in the International Protection of Global Cultural Goods' (2012) 23 Eur J Intl L 719, 721.

debate and obtain support for policies to mitigate climate change".[97] In other words, the iconic character of WHC sites can be seen as an asset for raising public concern and building up support to take preventive and precautionary measures in the context of climate change. The idea is that the loss of WHC sites is likely to be more tangible and immediate to people than meteorology and physical science. Second, there is the *comparative advantage* of WHC sites, *id est* the fact that they "can serve as laboratories where monitoring, mitigation and adaptation processes can be applied, tested and improved". The World Heritage Centre can thus promote, in cooperation with the States Parties, "the use of WHC properties in the activities of other conventions, international bodies and programmes working on climate change" because "(a)ctions taken at these iconic properties [...] can influence the adoption of good management practices elsewhere".[98]

To mobilise the policy value and the legal force of Cultural Heritage in the development of synergies with the climate change regime, the injection of Cultural Heritage concerns into climate change negotiations is crucial. One way to do so is through the adoption of a more proactive approach on the part of the WHC Committee. This could exercise pressure on the States and international organizations claiming leadership in matter of climate change demanding that the WHC be explicitly taken into consideration in the definition of mitigation strategies. Such an approach would be in line with the findings of the increased cooperation between WHC bodies and the international organizations working on climate change contained in the Policy Document[99] and with Article 13(7) WHC, which provides that the WHC Committee "shall cooperate with international and national governmental and non-governmental organizations having objectives similar to those of this Convention".

3. A Critical Appraisal

Although the establishment of synergies and linkages between the two areas of law under examination seems both desirable and necessary, this proactive hypothesis needs be considered in light of the following issues: (a) the functioning of the climate change regime; and (b) the nature and extent of States' obligations under the WHC.

---

97 UNESCO, 'Policy Document' (n 54) 9.

98 Ibid 4.

99 See n 91.

a) The Functioning of the Climate Change Regime

It is a fact that both the UNFCCC and the Kyoto Protocol (and the post-Kyoto regime) aim at combating anthropogenic global warming through cost-effective policies and measures. In this respect, Article 3(3) of the UNFCCC establishes that States Parties

> (S)hould take precautionary measures to anticipate, prevent or minimize the causes of climate change and mitigate its adverse effects. Where there are threats of serious or irreversible damage, lack of full scientific certainty should not be used as a reason for postponing such measures, taking into account that policies and measures to deal with climate change should be cost-effective so as to ensure global benefits at the lowest possible cost.

This approach has been heavily criticized. In particular, it has been maintained that the reductions agreed upon with the Kyoto Protocol (and the post-Kyoto regime) are inadequate and too modest to stabilize concentration of GHG emissions and to reverse current global warming and climate change trends. It has also been held that the Kyoto Protocol "can be considered as only a first and relatively small step towards stabilizing the climate".[100] The main problem is that the Protocol required only a small percentage of GHG emissions reduction – 5 per cent below 1990 levels between 2008 and 2012 – and only from industrialized States. This is an extremely modest requirement that, even if it were carefully fulfilled, could not curb the problem.[101] Indeed, to avoid temperature increases beyond 2°C above preindustrial temperatures, States should commit to greater reductions in GHG emissions than those called for by the Kyoto Protocol.[102]

The market-based mechanisms introduced by the Kyoto Protocol – Joint Implementation, CDM, Emissions Trading – have also attracted several criticisms. On the one hand, it has been argued that the solution to climate change should not lie in the dogma of market freedom and price mechanism. Many critics maintain that this approach is absurd because it encourages profit from pollution and subjects the development of green technologies to the approval of the market. These market-based mechanisms tend to reduce the natural world to a purely economic resource.[103] On the other hand, the mechanisms intro-

---

100 Scott Barrett, *Why Cooperate? The Incentive to Supply Global Public Goods* (OUP 2007) 91.

101 Leal-Arcas (n 78) 245.

102 Thorson (n 38) 265.

103 Adelman Sam, 'Rethinking Human Rights: The Impact of Climate Change on the Dominant Discourse' in Stephen Humphreys (ed) (n 1) 159, 164.

duced by the Kyoto Protocol obviate the fact that human life emerged from, and is dependent upon, the Earth's natural resource base and that human beings have a unique capacity to alter the environment.[104] In addition, by relying on market-based mechanisms, States and international organizations neglect the bitter lesson of the contemporary financial crisis and banking scandals that the market can no longer be trusted to self-regulate. In connection with this, it should be mentioned that chemical, wind, gas and hydro companies routinely abuse these systems by claiming emission reduction credits for projects that should not qualify.[105]

Accordingly, it appears that the abovementioned WHC Committee's deference to the UNFCCC system is misplaced.

From a political perspective, while the UNFCCC is optimally positioned to address the effects of climate change at the national and international levels, it has not fulfilled its promise to date. For this reason, the UNFCCC – and by extension the Kyoto Protocol – has been characterized as a quintessentially failed regime[106] because it results from consensus-driven and welfare-based solutions.[107] As a result, the climate change regime falls short of setting the necessary reductions in GHG emissions. Therefore, there is a real threat that much of the world's Cultural and Natural Heritage will be lost if the WHC Committee waits for the UNFCCC to 'solve' the climate change problem.

From a legal perspective, deference to the UNFCCC by the WHC Committee seems grounded on the principle *lex specialis derogat legi generali, id est* the principle that more detailed and specific norms ought to prevail against more general rules. In other words, the WHC Committee may have held the position that, as the UNFCCC was established to mitigate anthropogenic GHG emissions, States should use this regime also to address the problem of the impacts of climate change on Cultural Heritage.[108] However, the *lex specialis* principle is only relevant when legal norms clash. But in the case of the relationship between the UNFCCC and the WHC, it does not seem applicable because there is no language in the former treaty that evinces the intent to displace potentially

104 Dinah Shelton, 'Equitable Utilization of the Atmosphere: A Rights-Based Approach to Climate Change?' in Stephen Humphreys (ed) (n 1) 91, 115.
105 John Vidal, 'Billions Wasted on UN Climate Programme' *Guardian* (London, 26 May 2008).
106 Burns (n 47) 160.
107 Humphreys (n 5).
108 Burns (n 47) 158.

parallel mandates under other regimes to address climate change.[109] It can be presumed that the parties would have included such language if this was their intent. Moreover, it has been rightly pointed out that the *lex specialis* is the WHC because it contains the more detailed and specific norms in matters of protection of cultural assets in contrast to the more generalized mandate of the UNFCCC to protect the climate system.[110]

b) The Nature and Extent of States' Obligations under the WHC

The WHC defines the obligations of States Parties mainly in Articles 4, 5 and 6. Article 4 states:

> Each State Party to this Convention recognizes that the duty of ensuring the identification, protection, conservation, presentation and transmission to future generations of (cultural heritage sites) situated on its territory, belongs primarily to that State. It will do all it can to this end, to the utmost of its own resources and, where appropriate, with any international assistance and co-operation [...] which it may be able to obtain.

Thus States Parties accept the commitment to take all necessary actions to preserve any item of Cultural or Natural Heritage situated on their territory because their "deterioration or disappearance [...] constitutes a harmful impoverishment of the heritage of all the nations of the world".[111] Nevertheless, the WHC leaves it mostly to the States Parties to decide how to meet their obligations. Article 5 reads:

> To ensure that effective and active measures are taken for the protection, conservation and presentation of the cultural and natural heritage situated on its territory, each State Party [...] shall endeavor, in so far as possible, and as appropriate for each country: (a) to adopt a general policy which aims to give the cultural and natural heritage a function in the life of the community and to integrate the protection of that heritage into comprehensive planning programmes; (b) to set up within its territories, where such services do not exist, one or more services for the protection, conservation and presentation of the cultural and natural heritage with an appropriate staff and possessing the means to discharge their functions; (c) to develop scientific and technical studies

109 The *principle lex posterior derogat legi priori*, which is codified in Article 30 VCLT and which provides that where two treaties are concluded relating to the same subject matter the later treaty prevails, is inapplicable as between the WHC and the UNFCCC because they do not relate to the same subject matter. Catherine Redgwell, 'The World Heritage Convention and other Conventions Relating to the Protection of the Natural Heritage' in Francioni and Lenzerini (n 62) 377, 395.

110 Burns (n 47) 159–160.

111 WHC, preamble, second recital.

> and research and to work out such operating methods as will make the State capable of counteracting the dangers that threaten its cultural or natural heritage; (d) to take the appropriate legal, scientific, technical, administrative and financial measures necessary for the identification, protection, conservation, presentation and rehabilitation of this heritage; and (e) to foster the establishment or development of national or regional centres for training in the protection, conservation and presentation of the cultural and natural heritage and to encourage scientific research in this field.

Articles 4 and 5 are broad, potentially leaving much room for State Party discretion as to the exact nature of the respective responsibilities. The language used – "to the utmost of its own resources", "where appropriate", "endeavor", "in so far as possible" – indicates that these articles do not impose legally binding obligations.[112]

Article 6 is less discretionary. It provides that "it is the duty of the international community as a whole to co-operate" for the protection of WHC sites[113] and that the "States Parties undertake [...] to give their help in the identification, protection, conservation and presentation of the Cultural and Natural Heritage [...] if the States on whose territory it is situated so request".[114] Finally, it establishes that each "State Party [...] undertakes not to take any deliberate measures which might damage directly or indirectly the cultural and natural heritage [...] situated on the territory of other States Parties to this Convention".[115] The scope of Article 6(3) is both broad and narrow: it is narrow because only deliberate measures are covered; it is broad because it concerns any measure which may directly or indirectly damage Cultural Heritage.

All in all, Articles 4, 5 and 6 comprise the responsibility to cooperate to protect Cultural Heritage sites and to ensure that actions taken within a national territory do not cause damage or deterioration of items of Cultural Heritage situated in any other national territory.[116] Moreover, these provisions can be seen as the basis for States to address the causes and the potential and identified effects of climate change on WHC sites. Therefore, they certainly comprise the obligation to adopt site-specific mitigation and adaptation measures.[117] Furthermore, it

112 Thorson (n 38) 259.
113 Paragraph 1.
114 Paragraph 2.
115 Paragraph 3.
116 Thorson (n 38) 259.
117 See UNESCO, 'Report' Paragraphs 7, 77, 124–125; UNESCO, 'Strategy' Paragraphs 17–19; UNESCO, 'Policy Document' 4, 9. See also Thorson (n 38) 269.

can be argued that these norms oblige WHC States Parties to support the goals of the UNFCCC and the Kyoto Protocol.[118]

In the light of current climate change trends, however, the crucial question is whether Articles 4, 5 and 6 also entail an obligation for the States Parties to the WHC to make GHG emission reductions that go beyond the reductions called for by the Kyoto Protocol (and the post-Kyoto regime). Various reasons can be found in support of a positive answer to this question.

The first is that WHC obligations are independent of the obligations under the UNFCCC and the Kyoto Protocol.[119] Second, the WHC and the UNFCCC are not mutually exclusive. As said above, there is no language in the text of the UNFCCC that evinces the intent of the States Parties to exclude other legal regimes from addressing climate change when this is deemed necessary to effectuate the objectives of those regimes.[120] The third reason relates to the principle of no harm (or principle of prevention). This entails the duty of a State not to allow or tolerate any activity within its jurisdiction that may cause damage to the environment of other States or of areas beyond its national jurisdiction, unless the transboundary environmental impacts of this activity prove to be insignificant.[121] This principle, which belongs to customary law, was first enunciated in the *Trail Smelter* case[122] and was later incorporated in the 1972 Stockholm Declaration on the Human Environment,[123] the 1992 Rio Declaration on Environment and Development[124] and in the preamble of the UNFCCC. This principle was also taken up by the International Court of Justice in its Advisory Opinion on *Nuclear Weapons*[125] and in the judgments *Gabčikovo-Nagymaros*[126] and *Corfu Channel*.[127] Fourth, it can be argued that Articles 4, 5 and 6 entail an obligation to make GHG emission reductions beyond the limits established under the Kyoto Protocol (and the post-Kyoto regime) because, as demonstrated above, these re-

---

118 Gruber (n 13) 6.

119 Thorson (n 38) 267.

120 Burns (n 47) 159–160.

121 Francesco Francioni and Christine Bakker, 'The Evolution of the Global Environmental System: Trends and Prospects' (2013) 8/2013 Transworld Working Paper www.transworld-fp7.eu/?p=985 accessed 26 July 2013, 7.

122 *Trail Smelter Arbitration* (*United States v. Canada*) (Arbitration Tribunal) (1941) RIAA (1949) 1905.

123 Principle 21.

124 Principle 2.

125 *Legality of the Threat or Use of Nuclear Weapons* (Advisory Opinion) (1996) ICJ Rep 226.

126 *Gabčíkovo-Nagymaros Project* (*Hungary/Slovakia*) (1997) ICJ Rep 1.

127 *Corfu Channel Case* (*UK v. Albania*) (Merits) (1949) ICJ Rep 4.

ductions are inadequate. Additionally, the adoption of general GHG emissions reduction is necessary because site-specific mitigation and adaptation measures are inherently insufficient. Although the outstanding and fragile character of WHC properties justifies the adoption of site-level climate change strategies, any climate change intervention occurring within the boundaries of a WHC site cannot cope with the slow but devastating consequences of climate change on that site because GHG emissions normally occur outside WHC sites. Even if a State would impose a total ban on GHG emissions within the boundaries of a WHC site, this would continue to be threatened. Thus, site specific mitigation cannot realistically ameliorate the climate change threats to a WHC site in any meaningful way.[128] Moreover, there is a 'timescale mismatch' between mitigation measures and results. Even if every State were to implement drastic measures to reduce GHG emissions, it will be many decades before there are palpable effects because of the inertia of the climatic system.[129] In this respect, the strategy acknowledges that "(t)he benefit of mitigation at World Heritage sites is […] likely to be negligible on a quantitative basis".[130] In addition, it can be argued that WHC States Parties are under an obligation to make general GHG emission reductions because of the principle *pacta sunt servanda*. According to this, "(e)very treaty in force is binding upon the parties to it and must be performed by them in good faith".[131] Accordingly, States Parties are bound to implement the WHC through the adoption of measures effectively capable of realizing the purpose of this Convention.

However, the affirmation that the WHC contains a legally binding obligation for the States Parties to adopt GHG emission limits that are more compelling than those called for by the Kyoto Protocol is not uncontroversial. In effect, the adoption of this proactive interpretation unveils at least two problematic aspects.[132]

The first relates to the identification of the means to define the GHG emissions limits for States Parties to the WHC. Since the amendment of the text of the WHC is unlikely, one can consider two options: the adoption of an additional

128 Thorson (n 38) 270.

129 Burns (n 47) 156–157.

130 UNESCO, 'Strategy' (n 49) Paragraph 124.

131 Article 26 VCLT.

132 See Ottavio Quirico, 'Key Issues in the Relationship between the World Heritage Convention and Climate Change Regulation', in Silvia Borelli and Federico Lenzerini (eds), *Cultural Heritage, Cultural Rights, Cultural Diversity. New Developments in International Law* (Martinus Nijhoff Publishers 2012) 391, 406.

protocol or a further amendment of the Operational Guidelines. Apart from GHG emission limits, these new legal instruments should also provide measures of implementation, including sanctions in case of violations of mitigation measures. The second pertains to the *clausula rebus sic stantibus* codified in Article 62 VCLT. This provision makes it possible for a State to cite a "fundamental change of circumstances" as a ground for terminating, withdrawing from or suspending the operation of a treaty. Termination or withdrawal are possible only if (a) the circumstances that have changed were essential to the consenting of the parties to their legal obligations;[133] (b) the effect of the change is to radically transform the extent of the obligation in question;[134] (c) the obligation is unrelated to a boundary matter;[135] (d) the fundamental change is not caused by the defaulting State's previous breach.[136] Article 62 VCLT is relevant for the purposes of the present study because climate change was not perceived as a global threat when the WHC was adopted. In effect, the Convention does not cover the phenomenon of climate change. Accordingly, it can be reasonably argued that if the WHC were interpreted to impose an obligation to implement general GHG cuts, the States that currently oppose GHG emissions reduction targets would likely invoke Article 62 VCLT to terminate, withdraw from or suspend the operation of the WHC on the grounds that the absence of requirements of GHG emission reduction in the WHC constitutes an *essential basis of consent* to be bound by it. In other words, Article 62 would authorize the States Parties to the WHC to terminate, withdraw from or suspend the operation of this treaty because its implementation would differ from the original commitment. Needless to say, if the main GHG emitting States would decide to abandon the WHC in order to flout GHG, the mission of the WHC would be inevitably undermined.[137]

VI Conclusion

Climate change is a global problem that transcends territorial boundaries. Its impacts are now being felt and others cannot be halted due to the extent of historical and current emissions and the time lag between emissions and their effects on the climate. Thus far, the response of the international community has been disheartening, with UN negotiations on the reduction of GHG emissions

133 Article 62(1)(a) VCLT.
134 Article 62(1)(b) VCLT.
135 Article 62(2)(a) VCLT.
136 Article 62(2)(b) VCLT.
137 Burns (n 47) 161.

proceeding at glacial pace due to the stand-off between major economic powers.[138] Likewise, UNESCO bodies have failed to provide decisive responses. Moreover, this chapter has demonstrated that there is no constructive interaction between Cultural Heritage and climate change law. The international treaties that address the degradation of global climate conditions do not take account of the problem of the impacts of climate change on Cultural Heritage.

The main argument advanced here is that the evolution of the current state of affairs towards recognition of Cultural Heritage concerns by climate change law is both desirable and necessary. This complementarity would make it possible to emphasise that climate change is about losing the legacy – sites, monuments and social structures – handed over to us by our forbearers as a result of the damage that peoples are doing to nature. More importantly, as demonstrated by the studies endorsed by the WHC Committee, the building of a mutually supportive relationship between these areas of law would make it possible to raise public concern, build up support for preventive and precautionary measures and develop best practices and pilot projects in vulnerability assessments, monitoring, mitigation and adaptation processes.

However, although it would be germane that the legal instruments dealing with climate change were modified – so as to recognise that this phenomenon impacts Cultural Heritage – efforts should be first and foremost directed at addressing the shortcomings of the existing climate change regime. The above analysis indicates that otherwise any revision in the sense advocated above would be useless.

Needless to say, a meaningful reform of climate change law can be achieved only if States – especially industrialized and developing countries – accept to revise their agenda of unsustainable development. In effect, thus far national governments have proven to be unwilling to grasp the scale and urgency of the problem.[139] States have been more concerned with pursuing economic, political

138 Pilita Clark, 'Climate: Progress at Glacial Pace in UN Talks on Emissions' *Financial Times* (22 January 2013).

139 Of course, important differences persist among States. Certain States are determined to comply with climate change targets for the sake of the protection of global climate for present and future generations. The EU and its Member States have unilaterally decided to go beyond their emission reduction targets included in the Kyoto Protocol. With Decision No. 406/2009/EC of 23 April 2009 of the European Parliament and of the Council, the EU has set a target of cutting GHG emissions by 20 % in 2020 from 1990 levels. Other States do not accept binding commitments, as these would severely harm the national economy and their freedom to pursue

or military interests to prioritize the respect and the protection of global public goods. It is the continuing prevalence of State-centred perspectives that has hindered the development of legal obligations codifying the commitment of every State to engage in the promotion and protection of the global environment.[140]

In sum, today climate change can be seen as a potentially unique opportuni ty to develop more rational and egalitarian international governance structures and normative reforms not at variance with the principle of 'sustainable development'. As well known, the Brundtland Commission defined it as the "(d)evelopment that meets the needs of the present without compromising the ability of future generations to meet their own needs".[141] This principle emphasises that environmental protection is in everyone's interest and calls for the development of a global regulatory regime aiming at: (i) limiting or recasting the principle of State sovereignty; States should take account of the negative duty not to harm internationally recognized global public goods and should thus stop regarding polluting the atmosphere as a sovereign prerogative; State sovereignty should become a vehicle (not a barrier) for ensuring the protection and conservation of global public goods; humanity faces a threat that cannot be addressed if States do not accept to subordinate perceived national interests to supranational interests;[142] (ii) setting coherent and enforceable adaptation policies and mitigation targets; (iii) financing the development of new technologies that substitute fuels and preserve forests; (iv) redesigning international markets and trade; and (v) providing information about the impacts of climate change to the public at large.

---

development policies. The United States have been very reluctant to make any binding commitments at the multilateral level without any agreement on efforts to be made by newly industrialized and highly polluting states such as China and other developing countries (Francioni and Bakker (n 121) 11–15). In the US, the public debate has been heavily conditioned by the efforts of the oil and gas industry to debunk evidence of the causal link between burning of fossil fuels and climate change (Oriana Zill de Granados and AC Thompson, 'The Manipulation of Science' *PBS* (24 April 2007)).

140 Francesco Francioni, 'Realism, Utopia and the Future of International Law', in Antonio Cassese (ed), *Realizing Utopia. The Future of International Law* (OUP 2012) 442, 454– 456.

141 World Commission on Environment and Development, 'Our Common Future' (1987) Doc. A/42/427/Annex.

142 Adelman (n 103) 167.

Because climate change affects – and will continue to affect – the ecological systems that sustain life, not only the economic development of States, the opportunity to build international consensus around new international governance structures and normative reforms must be grasped now. Various recent alarming reports demonstrate that procrastination is not an option.[143] Sherwood Rowland, Nobel Prize laureate for his research on the effects of chlorofluorocarbon gases on the ozone layer, asked "(w)hat is the use of having developed a science well enough to make predictions if all we are willing to do is stand around and wait for them to come true"?[144]

Bibliography

Adelman, S, 'Rethinking Human Rights: The Impact of Climate Change on the Dominant Discourse' in Humphreys S (ed), *Human Rights and Climate Change* (CUP 2010) 159

AXA ART Group, 'AXA ART Presents Damages from Hurricane Sandy at Its Lounge During 2013 AIPAD Photography Show New York' (press release, 3 April 2013)

Barrett, S, *Why Cooperate? The Incentive to Supply Global Public Goods* (OUP 2007) Bodansky, D, 'The United Nations Framework Convention on Climate Change: A Commen tary' (1993) Yale J Intl L 451

Brooks, M, 'Scientists Finally Get Angry about Indifference to Climate Change' *Guardian* (5 July 2011)

Burns, W C G, 'Belt and Suspenders? The World Heritage Convention's Role in Confronting Climate Change' (2009) Rev European Community and Intl Environmental L 148.

Burns, W C G. and H M Osofsky, 'Overview: The Exigencies That Drive Potential Causes of Action for Climate Change', in Burns W C G and H M Osofsky (eds), *Adjudicating Climate Change. State, National, and International Approaches* (CUP 2009) 1

---

143 See Gail Whiteman, Chris Hope and Peter Wadhams, 'Vast Costs of Arctic Change' *Nature* (25 July 2013); and National Oceanic and Atmospheric Administration, 'Carbon Dioxide at NOAA's Mauna Loa Observatory Reaches New Milestone: Tops 400 ppm' Earth System Research Laboratory (10 May 2013) researchmatters.noaa.gov/ news/Pages/CarbonDioxideatMaunaLoareaches400ppm.aspx accessed 28 July 2013.

144 Michael Brooks, 'Scientists Finally Get Angry about Indifference to Climate Change' *Guardian* (5 July 2011).

Buzzini, G P and L Condorelli, 'Article 11: List of World Heritage Sites in Danger and Deletion of a Property from the World Heritage List', in Francioni F (edited with the assistance of F Lenzerini), *The 1972 World Heritage Convention. A Commentary* (OUP 2008) 175

Carducci, G, 'Articles 4–7: National and International Protection of the Cultural and Natural Heritage', in Francioni F (ed, with the assistance of F Lenzerini), *The 1972 World Heritage Convention. A Commentary* (OUP 2008) 103

Clark, P, 'Climate: Progress at Glacial Pace in UN Talks on Emissions' *Financial Times* (22 January 2013)

Francioni, F and C Bakker, 'The Evolution of the Global Environmental System: Trends and Prospects' (2013) Transworld Working Paper 8/2013, www.transworld-fp7.eu/?p=985 accessed 26 July 2013

Francioni, F, 'Public and Private in the International Protection of Global Cultural Goods' (2012) Eur J Intl L 719

Francioni, F, 'Realism, Utopia and the Future of International Law' in Cassese A (ed), *Realizing Utopia. The Future of International Law* (OUP 2012) 442

Francioni, F, 'Thirty Years On: Is the World Heritage Convention Ready for the 21st Century?' (2003) Italian Y Intl L 13

Gilbert, L, 'An Exodus from Red Hook' *The Art Newspaper* (26 April 2013)

Gruber, S, 'The Impact of Climate Change on Cultural Heritage Sites: Environmental Law and Adaptation' (2008) Sidney Law School-Legal Studies Research Paper 8/117, papers.ssrn.com/sol3/papers.cfm?abstract_id=1285741 accessed 26 July 2013

Hee-Eun, K, 'Changing Climate, Changing Culture: Adding the Climate Change Dimension to the Protection of Intangible Cultural Heritage' (2011) Intl J Cultural Property 259

Homer-Dixon, TF, 'Terror in the Weather Forecast' *New York Times* (24 April 2007)

Humphreys, S, 'Conceiving Justice: Articulating Common Causes in Distinct Regimes' in Humphreys S (ed), *Human Rights and Climate Change* (CUP 2010) 299

Humphreys, S, 'Introduction: Human Rights and Climate Change' in Humphreys S (ed), *Human Rights and Climate Change* (CUP 2010) 1

IPCC, Third Assessment Report, Synthesis Report, Working Group II: Impacts, Adaptation, and Vulnerability (Chapter 18.2) (2001) www.grida.no/publications/other/ipcc_tar/?src=/climate/ipcc_tar/wg2/644.htm accessed 27 July 2013

IPCC, Third Assessment Report, Synthesis Report, Working Group III: Mitigation (Appendix II: Glossary) (2001) www.grida.no/publications/other/ipcc_tar/?src=/climate/ipcc_ tar/wg3/466.htm accessed 27 July 2013

IPCC, Fourth Assessment Report: Climate Change (Cambridge University Press 2007)

Kaul, I, I Grunberg and M A Stern (eds), *Global Public Goods. International Co-operation in the 21st Century* (OUP 1999)

Ki Moon, B, 'A Climate Culprit in Darfur' *Washington Post* (16 June 2007)

Leal-Arcas, R, *Climate Change and International Trade* (Edward Elgar Publishing 2013)

Lenzerini, F, 'Articles 15–16: World Heritage Fund' in Francioni F (ed, with the assistance of F Lenzerini), *The 1972 World Heritage Convention. A Commentary* (OUP 2008) 269

Mutter, J C and K M Barnard, 'Climate Change, Evolution of Disasters and Inequality' in Humphreys S (ed), *Human Rights and Climate Change* (CUP 2010) 272

National Oceanic and Atmospheric Administration, 'Carbon Dioxide at NOAA's Mauna Loa Observatory Reaches New Milestone: Tops 400 ppm' (10 May 2013) researchmatters.noaa.gov/news/Pages/CarbonDioxideatMaunaLoareaches-400ppm.aspx accessed 28 July 2013

Quirico, O, 'Key Issues in the Relationship between the World Heritage Convention and Climate Change Regulation' in Borelli S and F Lenzerini (eds), *Cultural Heritage, Cultural Rights, Cultural Diversity. New Developments in International Law* (Martinus Nijhoff Publishers 2012) 391

Rayfuse, R and S V Scott, 'Mapping the Impact of Climate Change on International Law' in Rayfuse R and Scott S V (eds), *International Law in the Era of Climate Change* (Edward Elgar Publishing 2012) 3

Redgwell, C, 'The World Heritage Convention and other Conventions Relating to the Protection of the Natural Heritage' in Francioni F (ed, with the assistance of F Lenzerini), *The 1972 World Heritage Convention. A Commentary* (OUP 2008) 377

Shelton, D, 'Equitable Utilization of the Atmosphere: A Rights-Based Approach to Climate Change?' in Humphreys S (ed), *Human Rights and Climate Change* (CUP 2010) 91

Thorson, E J , 'The World Heritage Convention and Climate Change: The Case for a Climate Change Mitigation Strategy beyond the Kyoto Protocol' in Burns

W C G and H M Osofsky (eds), *Adjudicating Climate Change. State, National, and International Approaches* (CUP 2009) 255

Trachtman, J P, *The Future of International Law. Global Government* (CUP 2013)

UNESCO World Heritage Centre, 'Case Studies on Climate Change and World Heritage', available at: whc.unesco.org/en/activities/473 accessed 25 September 2013

UNESCO, Climate Change and World Heritage. Report on Predicting and Managing the Impacts of Climate Change on World Heritage and Strategy to Assist States Parties to Implement Appropriate Management Responses' World Heritage Reports 22 (1 May 2007)

Vidal, J, 'Billions Wasted on UN Climate Programme' *Guardian* (26 May 2008)

World Commission on Environment and Development, 'Our Common Future' (1987) Doc. A/42/427/Annex

World Heritage Centre, 'World Heritage Information Kit' (2008) whc.unesco.org/documents/publi_infokit_en.pdf accessed 27 January 2014)

Zill de Granados, O and A C Thompson, 'The Manipulation of Science' *PBS* (24 April 2007)

Birgitta Ringbeck*

World Cultural Heritage Sites and Climate Change: Management Issues

Since 2005[1] climate change has been on the World Heritage Committee's agenda as one of the greatest dangers to World Heritage sites. The committee responded to this emerging threat at its 29th session in 2006 by launching an initiative to assess the impacts of climate change on World Heritage.[2] Increased global temperature has been identified as just one of the consequences of the impacts of human activities on the climate equilibrium of the planet, with modifications of precipitation patterns, droughts, storminess, ocean temperature and acidification, sea level rise, etc. Aside from physical threats on natural and cultural sites of outstanding universal value, climate change will impact intangible heritage. Strategies and appropriate management responses are requested, especially for Cultural Heritage where the level of awareness and research is not as high as it is for Natural Heritage.

Also, European and German cultural sites are affected by climate change even though it is no not always clear, without ambiguity. For instance, the recent floods sweeping across Central and Eastern Europe, in June 2013, are indirectly tracked back to climate change and directly to other man made interventions such as canalization of rivers and damage of riparian zones. Often it is a combination of causes. Water related hazards account for 90 per cent of all disasters, and their frequency and intensity are rising. The floods affecting Eastern and Central European sites illustrate that cultural, and especially urban, heritage is particularly at risk and needs to be better protected against such events, especially considering its crucial role for the economy of local communities and their overall well-being. More effective coordination policies for heritage conservation and flood risk prevention at the national and local levels require the full

---

* Federal Foreign Office Germany (Auswärtiges Amt), department of multilateral cultural and media policy; member of the UNESCO World Heritage Committee.

1 UNESCO, World Heritage Committee Decision 29 COM 7B.a (9 September 2005) WHC-05/29.COM/22.

2 UNESCO, World Heritage Committee Decision 30 COM 7.1 (23 August 2006) WHC06/30.COM/19.

integration of heritage concerns into risk strategies and programs, before any hazards occur.

Management plans for World Heritage sites are key-tools in the effective stewardship of World Heritage sites under threat from climate change. As a regularly updated working document, such a plan should describe actions and include measures in response to climate change, *id est* education and traditional skills, monitoring and maintenance, research to support national/regional decision-making planning for emergency preparedness, re-evaluation of management priorities and training on the various problems. Moreover, risk and vulnerability maps of the site, which overlay flood data and heritage site locations, so that an overview of the risks can be obtained and detailed adaption strategies developed, are important.

By now, nearly every management plan for a German World Heritage site identifies threats and describes measures for preventive protection. Flood defence and action plans have been developed for most of the German sites inscribed on the World Heritage List and on the tentative list. They describe the area, indicate the fold areas, quantify possible damage and suggest general and local measures.

Floodplains based on 100to 500-year-experiences and empirical data are important for risk preparedness. Traditional knowledge and long-time-experience illustrate that unforeseeable natural disasters are not a new natural phenomenon, they occur from time to time. For generations, one of the most effective measures to protect a building against floods was the careful examination of the building's foundation and to take care for sufficient flood zones. The floodplain (*Figure 1*) for the Cultural Heritage site *Carolingian Westwork and Civitas Corvey* documents that the Benedictine monkhood were fully aware of such a hazard; in 822 Corvey abbey was founded exactly at the point in a curve of the River Weser which, for over 1200 years, has never been affected by floods. A 100-year water level fluctuation affects only select parts of the open spaces within the monastery grounds, surrounded by walls, in the area of the physical structure of Corvey.

Figure 1: Map of Floodplains for the Cultural Heritage site Carolingian Westwork and Civitas Corvey (Town of Höxter)

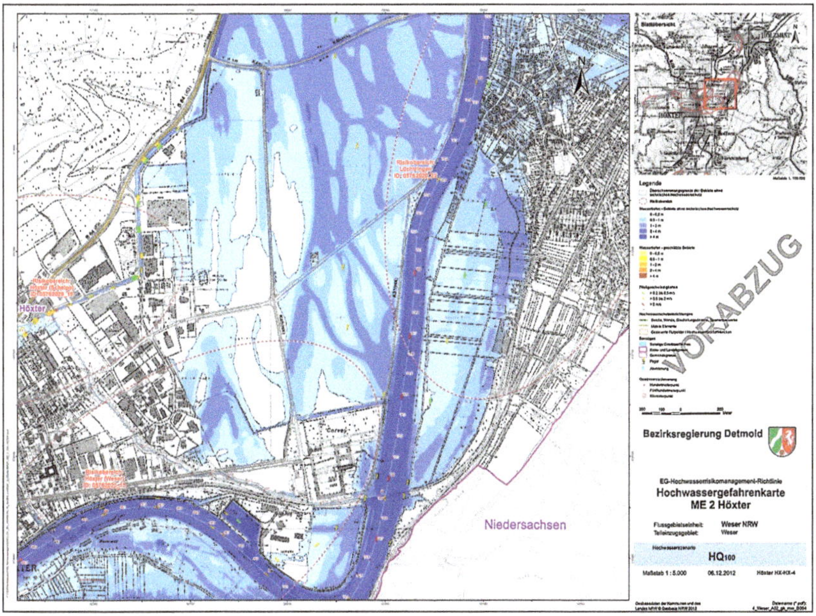

In order to avoid negative impacts of flooding and to minimize damages, alarm and operation plans, content attendance and resource availability as well as suitable preventive measures like raising the level of dikes and mobile flood-walls are necessary. In the World Heritage city *Old Town of Regensburg and Stadtamhof* for example, an interdisciplinary competition was held in 2003 for technical and design solutions for flood protection in the municipal area. The result focused primarily on stationary and mobile solutions as well as different combinations of these. In and around the area of the less effected Old Town river-bank, man-made mobile metal elements (*Figure 2*) are planned. In other areas, both mobile barrier sections, as well as stretches with combinations of stationary plinth walls and mobile barrier systems, are planned. In realizing the flooding protection system, special attention had to be paid to ensure that these stationary elements would not degrade the visual qualities.

Figure 2: Mobile Flood-Walls in the Old Town of Regensburg and Stadtamhof

Flood protection systems have to be compatible with the integrity of World Heritage sites. The buffer zone of the nominated site *Speicherstadt and Kontorhaus district with Chilehaus* (Hamburg) is a good example of the reconciling of flood and heritage protection requirements. Existing flood defenses along the Customs Canal are designed and integrated as viewpoints into the canal structure (*Figure 3*).

Figure 3: Existing flood defences along the Customs Canal and their utilisation as viewing points

Cases like these demonstrate that climate change is a global threat which needs answers and management responses on the state, city, or even neighborhood level. Beside preventive protection measures for every site, a general policy is necessary to address and minimize the impact of climate change. In Germany, this issue has been taken very seriously; especially the promotion to expand renewable energy has been increased by the national and federal governments in the last years. Appropriate laws and regulations ensure a guaranteed feed-in remuneration for electricity produced by solar panels, photovoltaic systems, biogas plants and wind mills as well as the connection to and distribution through the power grid. Building insulation is subsidized by the state. The political goal is to reduce the emissions caused by fossil fuels in order to slow down the global warming and to cover 80 per cent of the demand by energy produced from renewable energies in 2050. The measures already taken show their effects: Investments made in renewable energy generation plants pay off in the short and medium term due to state support and fiscal privileges. The awareness building has been successful; state subsidy programmes have contributed significantly to convince the public. First important steps for the energy turnaround and on the way from the fossil fuel and nuclear age to the solar and efficient energy age are done.

There is probably no country in Europe in which the using of renewable energies is that visible as in Germany. In some regions and historic cities wind farms, photovoltaic and biogas plants as well as insulated buildings are dominant elements. It sometimes looks as if the “Land der Dichter und Denker” has been turned into a “Land der Dichter and Dämmer”. This ironical remark points to the other side of the coin; renewable energy generation plants could have serious impacts on World Heritage sites especially in regard to their integrity and visual qualities. In the reflection reports on the trends of conservation – annually presented to the World Heritage Committee[3] – wind farms have long since been recognized as possible threats for the integrity of a site due to the need to be on exposed sites to catch the wind. Their impact can be highly detrimental in visual terms to the setting of World Heritage properties, particularly in flat open landscapes and mountain ridges where they can disturb long views, panoramas and silhouettes. Solar panels and photovoltaic systems could change completely the roof landscapes of historic cities and traditional villages as well as agricultural landscapes. Building insulation meant to destroy the surface of historic façades, in the end the character of a complete traditional ensemble could get lost.

---

3 See UNESCO, World Heritage Committee Decisions, always Number 7c on the list of documents.

Heritage assets and conservation requirements concur often with measures and projects for saving energy and producing renewable energy. A balancing of different public and private interests, as to slow down the global warming and to reduce energy costs, is necessary. The way ahead is to look for alternative solutions. Heritage Impact Assessment based on a clear articulation of the attributes of Outstanding Universal Value (OUV) and sound indisputable topographical data is essential for defining the potential visual impacts of windmills and other renewable energy generation plants in properties of outstanding universal value and their buffer zones.

In general, not enough attention is given to the fact that use and reuse of monuments is to be seen as ecologically sound and economically advantageous because of the potential energy sources in buildings, the reduction of land consumption for construction, and the know-how transfer of traditional skills. Subsidy programmes are very narrow, focused on renewable energy generation; built heritage is not as appreciated as non-renewable resources. In a nutshell, it may be concluded that to a large extent management issues with regard to World Cultural Heritage and climate change remain unresolved or even still untackled.

Roland Bernecker*

Concluding Remarks

The impact of climate change on our Cultural Heritage is a fact. 72 per cent of the State Parties to the World Heritage Convention[1] report damages that can be linked to the effects of climate change, 125 sites in 59 countries are affected. Extreme events related to climate change will increase in frequency and in intensity. In addition, we have to consider slow events like changes in corrosive processes, the effects of which are less obvious. The threat is not just one to the buildings and the material side of things: losses and damages we incur concern socio-economic patterns as well. The loss of memory and of historic wisdom in dealing with our cultural and natural resources may lead to the disruption of peaceful coexistence.

The rich and diverse presentations of the conference "Climate Change as a Threat to Peace: Impacts on Cultural Heritage and Cultural Diversity" have shown that culture is a crucial, yet dramatically neglected dimension of sustainability. More research will be necessary, especially in order to assess local impacts of climate change, to analyse risks and to better capture the relevance of socio-economic factors. At the conference, a concrete proposal was made to establish a World Monitoring Report on the state of World Heritage Sites, similar to UNESCO's Global Monitoring Report on Education. This would help not only to collect information, which is already available on a large scale, but also to process this information adequately. At present, this does not happen for lack of funds. Such a report could also help to use the prestigious World Heritage Sites to better communicate the destructive effects of climate change.

At the same time, we have to reconsider the concepts of prevention, adaptation and monitoring with regard to the impact of climate change on culture and peace. We have to broaden our view and rework the concepts we use in describing and analysing Cultural Heritage. What does *preservation and safeguarding of Cultural Heritage*, what does *change*, what does *management* mean? *Climate change* itself has to be put in the more comprehensive frame of what might be

* Secretary-General of the German Commission for UNESCO (DUK).

1 Convention concerning the Protection of the World Cultural and Natural Heritage; adopted 16 November 1972, entered into force 17 December 1975, 1037 UNTS 151.

better captured in the expression *global change*. We are not only facing more extreme weather conditions. The dimension of a changing climate is to be seen in relation to significant demographic developments and the important effects of a highly dynamic global urbanisation. An interdisciplinary and holistic approach is necessary when we reassess the conceptual frame of our analysis and reconsider the function of heritage in the context of global changes.

A major issue in this regard is the close link between natural and cultural resources. Both categories of heritage are integrated in the World Heritage Convention of 1972, but not until 1992 was the new concept of cultural landscapes as sites representing both dimensions officially introduced. We are still at the beginning of learning to understand interdependencies.

Moreover, throughout the course of the conference an important issue repeatedly emerged: Cultural Heritage and cultural diversity are an important factor in developing more sustainable societies. In this context, the notion of *resilience* appeared as crucial. It means the capacity to resist to, absorb and recover from effects of hazards. In view of increasing threats, resilience is becoming more important. We must shift from a paradigm of simple growth and efficiency to better deal with risks and uncertainty. Cultural diversity is a recognised principle in dealing with uncertainty and it further increases resilience. The fragility of monocultural patterns is well known. Again, viable solutions have to be developed in the specific context at the local level. Urban resilience, the development of *smart cities* with intelligent use of ICTs, is an inspiring concept in the perspective of enhancing the resilience of our societies.

We are suffering today from a technology bias in the search for solutions for the problems we are facing. We tend to increase the complexity and specificity of our technical tools in order to solve problems. Cultural resources are neglected, as is the need to redefine some of our basic problems in a cultural way. Cultural heritage in its different dimensions is reflecting human experience in the permanent effort to build resilient communities. To care for heritage means to value our long-term collective experience. The World Heritage Convention is not dealing only with sites, but also with values.

One layer of the day's discussions addressed the legal issues linked to the preservation of Cultural Heritage. With 190 State Parties, UNESCO's World Heritage Convention is universally ratified. In Article 4, the convention clearly defines the obligation of State Parties which I consider worthwhile reproducing here *in extenso*:

> Each State Party to this Convention recognises that the duty of ensuring the identification, protection, conservation, presentation and transmission to future generations of the cultural and natural heritage referred to in Articles 1 and 2 and situated on its

> territory, belongs primarily to that State. It will do all it can to this end, to the utmost of its own resources and, where appropriate, with any international assistance and co-operation, in particular, financial, artistic, scientific and technical, which it may be able to obtain.

The question is how the obligation laid down in this provision could be enforced with regard to the destructive effects of climate change. Do we really, besides the ethical pressure, dispose of some kind of legal leverage to bring governments to action? The World Heritage Convention itself does not address the issue of climate change directly. The only practical way to add this dimension would be a revision of the Operational Guidelines where the threats posed by climate change and necessary action could be included. The same is valid for other cultural conventions as for example the Intangible Heritage Convention[2] which is totally unequipped with regard to climate change.

Other international treaties should be analysed for points of convergence and supportiveness in providing a legal hold for advancing concrete requests. The current negotiation of newly defined Sustainable Development Goals could be an entry point to better integrate the issue in the international agenda. Seen that the impact of climate change on Cultural Heritage does constitute, in a broader perspective, a possible threat to peace, the proposal was made to bring this issue to the Security Council of the UN. The Security Council would be in a position to declare its own competence in this matter and to take immediate action.

The more basic question was voiced of if and how the challenges posed by climate change present an opportunity to limit State sovereignty in the view of better securing global public goods.

When we consider the destructive impact of anthropogenic climate change on World Heritage Sites, we run into a highly symbolic paradox: Cultural sites of outstanding universal value, globally recognised as the greatest achievements of human creativity and as the most treasured expressions of cultural identities, are threatened by the effects of human inventiveness in developing unsustainable lifestyles. While we inscribe outstanding natural sites in UNESCO's World Heritage List, thus formally accrediting their critical value as biological and aesthetic resources for humanity, these sites are threatened by a feverish exploitation of the natural resources of our planet. Anthropogenic climate change seems to be a result of the same cultural disposition, which allowed humanity to construct Cultural Heritage of outstanding universal value. Today, our

2 Convention for the Safeguarding of the Intangible Cultural Heritage; adopted 17 October 2003, entered into force 20 April 2006, 2368 UNTS 3.

inventiveness and our hunger for improvement not only compromise our future, but consume our past.

We have to better explain why – when addressing the challenges of climate change and global peace – Cultural Heritage is a fundamental issue. The concern of heritage conservation is most evidently losing ground. In my view, we have to be more specific when we talk about Cultural Heritage, focusing more on the values we need to preserve. We have to be more concrete and more compelling when we ask for maintaining awareness of the human dimension, awareness of who we were in the past, who we are today, and what the world will look like in the future.

In reflecting upon the loss of bonds with our heritage, we have to become more serious. The World Heritage Convention is about values. These values are the essence of how we live, what is vital for us. My impression is that in the relentless race for an ever-increasing effectiveness, we are becoming inattentive to our past, negligent of the interdependencies with our natural environment and that we too zealously stress the human substance of our societies. This will make us more vulnerable.

It is interesting to note that *vulnerability* is a central feature in understanding the World Heritage Convention. In the 1960s, humanity was offered the opportunity to learn to consider its presence on earth as a somewhat surprising, at least fragile, contingency. On July 20th 1969, an estimated 600 million people followed the landing on the moon of Apollo 11 on TV. For the first time in history, a human being stood on firm ground outside the geosphere and had an inverted look on it. To us humans living here, the earth has become the epistemological angle of all cosmic events. Standing on the moon with Neil Armstrong, we had to perceive the blue planet as an object, shining with beauty and life, but surrounded by an infinite black space which seemed unrelated to human destiny, indifferent to our efforts and hopes. In the year 1968, Richard Buckminster Fuller, an American architect and futurist, published the legendary book, *Operating Manual for Spaceship Earth*, in which he sets forth the metaphor of the earth as a mechanical vehicle that requires constant maintenance and that will cease to function if we do not keep it in good order. "We are all astronauts", says Fuller, and there is no emergency exit from our spaceship.

In line with Buckminster Fuller's metaphor, the international community started in the same years to worry in more concrete terms about the maintenance of this vehicle. On June 5 to 16, 1972, the first United Nations Summit on the Human Environment took place in Stockholm. It was the summit that led to the establishment of the United Nations Environmental Programme. The perception that we had to take care of planet earth had become an issue of high intergovernmental relevance. It is for this Stockholm summit that finally natural sites were included in the World Heritage Convention, adopted some five months later in Paris.

In addition to this emerging change of perspective on our terrestrial conditions, the possibility of a comprehensive nuclear destruction had become a realistic scenario. Some weeks ago I visited the former German government bunker in Ahrweiler near Bonn, constructed between 1960 and 1972, in order to allow a continuation of central elements of the political administration for at least 30 days after a nuclear attack. The bunker was fully functional until 1997 and is now a heritage site. Not only the approximate cost of 1.5 billion € for this impressive construction indicates how seriously the highest levels of the political establishment considered the probability of a nuclear war, it is also the detailed thoroughness of its implementation, its technological ambition and the bureaucratic soberness of its functionality that still today create a strongly embarrassing feeling of how close humanity seemed to have come to the moment the bunker was build for.

The inventory of human achievements we collect in the prestigious World Heritage List was adopted the same year the construction of the bunker was finished. One might see the World Heritage Convention as the collective response to these developments that confronted us as cruelly as never before with the fundamental vulnerability of the human condition. The basic idea, the starting point of the World Heritage Convention, is the concept of "one humanity", of the "unity of human values" as it is put in the Venice Charter of 1964. The concept of a common heritage of humanity is indeed the most sublime reply to the undefeatable vulnerability and fragility of human life on earth. Climate change is adding a new dimension to the vulnerability of the ecosystem we depend upon.

The World Heritage Convention reframes the competitive set of national cultural pride into the global perspective of shared universal values, which help us to find common responses to growing global challenges. Not to loose these values is a necessary contribution to peace.

Dresdner Schriften zu Recht und Politik der Vereinten Nationen
Papers on Law and Policy of the United Nations

Herausgegeben von Sabine von Schorlemer

Band 1 Sabine von Schorlemer (Hrsg.): „Wir, die Völker (...)" – Strukturwandel in der Weltorganisation – Konferenzband. Konferenz aus Anlass des 60-jährigen Bestehens der Vereinten Nationen vom 27.–29. Oktober 2005 in Dresden in Kooperation mit dem Zentrum für Internationale Studien (ZIS) und der Deutschen Gesellschaft für die Vereinten Nationen (DGVN). 2006.

Band 2 Philipp von Hülsen: Die Vereinten Nationen als Gesetzgeber im Kosovo. Rechtmäßigkeitsanforderungen an Rechtssetzungsakte der UNMIK am Beispiel der Unternehmensprivatisierung durch die Kosovo Trust Agency. 2007.

Band 3 Sabine von Schorlemer (Hrsg.): Die Vereinten Nationen und neuere Entwicklungen der Frauenrechte. 2007.

Band 4 Tina Miller: Die Frage der Rückkehr palästinensischer Flüchtlinge, unter Berücksichtigung der Lösungsansätze der Vereinten Nationen. 2007.

Band 5 Isabel Hörtreiter: Die Vereinten Nationen und Wirtschaftsunternehmen – zwischen Kooperation und Kontrolle. Steuerungsformen zur Stärkung menschenrechtlicher Unternehmensverantwortung unter dem Dach der Vereinten Nationen. 2007.

Band 6 Ariane Wewers: Die Pflicht der EU-Partner zur Koordinierung in internationalen Organisationen und auf internationalen Konferenzen (Art. 19 EUV). Die Praxis der GASP dargestellt am Beispiel der Vereinten Nationen. 2009.

Band 7 Niklas Keller: Die „Principles on Housing and Property Restitution for Refugees and Displaced Persons" der Vereinten Nationen. Die Lösung der Wohnraum- und Eigentumsproblematik bei der Rückkehr von Flüchtlingen und Displaced Persons. 2009.

Band 8 Martin Ramsperger: Die Terrorismusbekämpfung im Rahmen der Europäischen Sicherheits- und Verteidigungspolitik (ESVP). 2009.

Band 9 Sabine von Schorlemer: Kindersoldaten und bewaffneter Konflikt. Nukleus eines umfassenden Schutzregimes der Vereinten Nationen. 2009.

Band 10 Lauri Philipp Rothfritz: Die Konvention der Vereinten Nationen zum Schutz der Rechte von Menschen mit Behinderungen. Eine Analyse unter Bezugnahme auf die deutsche und europäische Rechtsebene. 2010.

Band 11 Arlett Rassel: Strafgerichtsbarkeit über Angehörige der Friedenstruppen in UN-geführten Missionen. Unter besonderer Berücksichtigung von Sexualstraftaten. 2010.

Band 12 Dirk Monsau: Vereinte Nationen und Korruptionsbekämpfung. 2010.

Band 13 Jana Hertwig: Die Europäische Union und die Bekämpfung der Verbreitung von Massenvernichtungswaffen. Theorie und Praxis der europäischen Nichtverbreitungsstrategie. 2010.

Band 14 Jana Hertwig / Sylvia Maus / Almut Meyer zu Schwabedissen / Matthias Schuler (eds.): Global Risks. Constructing World Order through Law, Politics and Economics. 2010.

Band 15 Sabine von Schorlemer / Elena Schulte-Herbrüggen (Hrsg.): 1989–2009: 20 Jahre UN-Kinderrechtskonvention. Erfahrungen und Perspektiven. 2010.

Band 16 Philip Michael Winkler: Technologietransfer im Recht der UNESCO. 2012.

Band 17 Martin Gerner: Lead Agency. UNESCO´s Global Leadership and Co-ordination Role for the United Nations Decade of Education for Sustainable Development (2005–2014). 2013.

Band 18 Heidrun Groß: Kulturelle Vielfalt und Urheberrecht. Urheberrecht als Instrument im Rahmen des UNESCO-Übereinkommens zum Schutz und zur Förderung der Vielfalt kultureller Ausdrucksformen. 2013.

Band 19 Sabine von Schorlemer/Sylvia Maus (eds.): Climate Change as a Threat to Peace. Impacts on Cultural Heritage and Cultural Diversity. 2014.

www.peterlang.com

www.ingramcontent.com/pod-product-compliance
Ingram Content Group UK Ltd.
Pitfield, Milton Keynes, MK11 3LW, UK
UKHW021829100926
8556IPUK00004B/245

9 783631 662236